FUNDAMENTAL ASTRONOMY AND SOLAR SYSTEM DYNAMICS

Invited Papers Honoring Prof. Walter Fricke on the Occasion of His 70th Birthday

Edited by

R. L. DUNCOMBE, Austin
J. H. LIESKE, Pasadena
and
P. K. SEIDELMANN, Washington

Reprinted from Celestial Mechanics, Vol. 37 No. 3 (1985)

D. Reidel Publishing Company / Dordrecht / Boston

Library of Congress Cataloging-in-Publication Data

CIP

CIP-data appear on separate card.

ISBN 978-94-010-8584-7 ISBN 978-94-009-4688-0 (eBook)
DOI 10.1007/978-94-009-4688-0

Published by D. Reidel Publishing Company,
P.O. Box 17, 3300 AA Dordrecht, Holland.

Sold and distributed in the U.S.A. and Canada
by Kluwer Academic Publishers,
190 Old Derby Street, Hingham, MA 02043, U.S.A.

In all other countries, sold and distributed
by Kluwer Academic Publishers Group,
P.O. Box 322, 3300 AH Dordrecht, Holland.

Prof. Dr. WALTER FRICKE

TABLE OF CONTENTS

FUNDAMENTAL ASTRONOMY AND SOLAR SYSTEM DYNAMICS

PREFACE

"Fundamental Astronomy and Solar System Dynamics", a program of invited papers honoring Professor Walter Fricke, who for thirty years has been Director of the Astronomisches Rechen Institut in Heidelberg, was held at the Thompson Conference Center of the University of Texas at Austin on Wednesday 27 March 1985 on the occasion of his seventieth birthday and retirement as Director of ARI. Professor Fricke's contributions to astronomy encompass the areas of galactic dynamics, radial velocities, stellar statistics, the fundamental reference system and the constant of precession.

Participants were welcomed to the University of Texas by Professor J. Parker Lamb, Chairman of the Department of Aerospace Engineering and Engineering Mechanics. The presentations ranged from discussions of astrometric problems concerned with the reference system, the constant of precession, major and minor planet observations, planetary ephemerides and lunar and satellite laser ranging, to a study of disc galaxies in massive halos. The program concluded with a review of Professor Fricke's career. The three sessions were chaired by Victor G. Szebehely, Carol A. Williams and Jay H. Lieske.

The participants in this meeting, and in the Division on Dynamical Astronomy meeting that followed, were happy that Professor Fricke was able to attend. His presence at these meetings, as well as his thoughtful comments, were greatly appreciated. We are pleased to acknowledge the support of the Center for Space Research, the Department of Astronomy and the Department of Aerospace Engineering and Engineering Mechanics of the University of Texas at Austin.

RLD
JHL
PKS

Welcome

J. Parker Lamb

Department of Aerospace Engineering
and Engineering Mechanics

Ladies and Gentlemen:

On behalf of the Department of Aerospace Engineering and Engineering Mechanics I am pleased to welcome you to our campus which, I think, is one of the most modern in the U. S. We are especially proud of this conference facility which allows us to host many technical meetings each year in an ideal setting.

Since the early days of the U. S. Space Program, the Department of Aerospace Engineering and Engineering Mechanics has played an important role in education and research within the fields of celestial mechanics and artificial satellite applications. We are extremely proud that, over the past few years, our program - through the efforts of our distinguished faculty - has gained worldwide recognition for excellence. Thus, many of our faculty are major participants in the technical activities of the American Astronomical Society. We are also especially pleased to serve as host for this meeting at which you recognize the distinguished career of Professor Walter Fricke.

Once again, let me indicate how happy we are that you came to our campus and wish you an enjoyable meeting.

FUNDAMENTAL REFERENCE SYSTEMS; PAST, PRESENT, AND FUTURE

P. K. Seidelmann
U. S. Naval Observatory

ABSTRACT. The present fundamental reference system is defined by the FK5 Fundamental Reference Catalog and the associated IAU Resolutions of 1976 through 1982. The past is defined to be what preceded the present fundamental reference system and future is defined to be what ever will come after it. The resulting characteristics of the different fundamental reference systems are considered from the aspects of the observational data, the computational methods, and the underlying theories. Having developed an understanding of these operational fundamental reference systems, the desired properties of the ideal fundamental reference system are presented. Finally, some questions concerning the future fundamental reference system are presented.

INTRODUCTION

This meeting honors Professor Dr. Walter Fricke, who has done more than anyone else to ensure the accuracy and quality of the FK5 Fundamental Catalog. His efforts will influence and affect the fundamental reference system of the future. It is probably presumptuous of anyone to deliver a talk on the Fundamental Reference System in the presence of Walter Fricke. Certainly, we would all learn more if he spoke on the subject rather than I. However, it is not customary, nor proper, to ask the person of honor to do the work of preparing a paper and it is customary to include papers on the principal areas of interest of the honoree. Thus, it is a privilege for me to have the opportunity to speak on the subject of the Fundamental Reference System at a meeting in honor of Walter Fricke. Perhaps, I am sufficiently detached to have an objective perspective concerning the fundamental reference systems.

I think it is necessary to examine both the past and the present to obtain a prospective for the future. I am defining the present as the FK5 System as defined by the IAU resolutions passed in 1976, 1979 and 1982 and as introduced for 1984. The basis of this system is documented in the Supplement to the 1984 Astronomical Almanac. The past is the Fundamental Reference System as it was prior to 1984. The future is what will follow after the FK5, either as a replacement or a

Celestial Mechanics **37** (1985) 199–207.

modification of the FK5. If you seek a definition of the Fundamental Reference System, you quickly learn that this is a difficult problem. Mueller, (1969) states "An explicit definition of a fundamental system beyond the somewhat trivial statement that it is 'all the information contained in the positions and proper motions in a given fundamental catalog' has never been given, and the exact nature of a fundamental system must therefore be defined operationally". Eichhorn, (1974) states "a fundamental catalog defines a fundamental system." Walter Fricke, 1972 and 1979, states "The reference system defined by fundamental catalogues is that of the Earth's equator as the fundamental plane from which declinations are reckoned and the vernal equinox as zero point of right ascensions". Woolard and Clemence (1966) state that "the system of geocentric coordinates which the equator and equinox of date define on the celestial sphere, and the measures of sidereal and mean solar time which the equinox and the fictitious mean sun determine, constitute the fundamental reference system of observational astronomy."

It is my opinion that a complete definition of the fundamental reference system includes the time scales, the underlying theories, and the means of achieving the reference system. So it is my intent to include all these aspects in this review of the fundamental reference system. Perhaps the question needs to be raised immediately, why do we need a fundamental reference system? The simple answer is "to avoid chaos." The more practical answer is that all observations, missions and analyses are dependent upon some sort of reference system. In general, it is assumed that the errors in the reference system are sufficiently small that they do not affect the functions that are dependent upon the reference system. Certainly, that is one of the goals for the fundamental reference system.

OBSERVATIONAL BASIS

In the past, the observational basis for the fundamental reference system was entirely optical observations. Due to the uneven distribution of the observatories many more observations were made from the northern hemisphere than from the southern hemisphere. Therefore, the star positions and proper motions were more accurate in the northern hemisphere and significant systematic deviations existed in the fundamental reference system as a function of declination. There is also a significant variation in the accuracy of the observations as a function of time. The more recent observations are much more accurate. A limited number of the stars had known parallaxes and radial velocities; in almost all cases these were tabulated separately and had different epochs than the positional data.

The FK5 Catalog will be primarily based on optical observations. There has been a concerted effort to improve the observational data for the southern hemisphere, but it will still not be equivalent to the northern hemisphere. Also the proper motions for the southern hemisphere stars will in many cases be poorer than for the northern hemisphere. The origin and the reference plane for the FK5 is based on solar system ephemerides, which are based on optical, radar, laser-

the calculations were to be done by a calculator and it was desirable to place the burden of the calculation on the centralized computing organizations which would provide tabular ephemerides and apparent places. Thus, methods of convenience, such as the use of day numbers and the inclusion of E-terms in the apparent places, were desirable.

The present methods of computation are based on the availability of computers and avoidance of approximations for precise computations. The limitations of today are not computation capability or speed, but involve limitations on data transfer, incompatibility of programs on different computers, and variations in the methods of computation which were not significant in the past.

The future should see compatibility of computer programming, such that standard programs could be transferred routinely from one computer to the other, data transfers accomplished without difficulty, and computations performed directly at the telescope by means of small computers. The limitations on accuracies will be those imposed by the word length of the computer and more significantly by the physical limitations of the observational accuracy.

APPROXIMATIONS

In the past, the approximations in computing astronomical positions were based on the computing speed and the accuracy requirements of the observations. The celestial sphere was an adequate model and approximations were introduced without documentation or were not recognized as approximations. At the present time the interrelationships between time, the reference catalog, and the observational and reduction procedures become the principal limit of the accuracy when dealing with older observations. It is virtually impossible to correct the observations for the inaccuracies in the constants used in the reduction procedure, for systematic deviations in the reference catalog, or for an inaccurate theory of nutation or proper motions.

It is hoped that in the future the approximations will not be embedded in the computations, so that improved values will permit improved computations and improvements of observations. The real limitations will be based on the limits of our knowledge, rather than the approximations of convenience.

ACCURACIES

The accuracy of the fundamental reference system in the past was approximately one second of arc. The accuracies were consistent with the optical observations, which are less accurate as you go back in time. The deficiencies in the Fundamental Reference System were recognized in the recent past. The accuracy at the present is approximately 0.05-0.10 second of arc and is due to the limitations of the observational methods and the interrelationships between observations. Thus, the fundamental reference system is intertwined in

ranging, and spacecraft data. There are radio observations of greater accuracy than the optical observations, however these observations cannot be defined with respect to the fundamental reference frame with the full accuracy of the observations. Also, in many cases the radio "point" sources, originally unresolved, were later resolved and there is the possibility that the sources may change structure, form, and location.

The future offers independent observational data in many wavelengths. It is hoped that the future observations will not have been subjected to irreversible processing. The basic observational data will be retained in machine-readable form, so that improvements in the methods of processing might result in improved observational accuracy. A complete record will be maintained of the corrections which have been applied to these observations. The future observations will be directly tied to an independent time system based on atomic time. The Hipparcos satellite, the Space Telescope and a truly astrometric satellite, offer the prospects for observations distributed over the entire sky. If the astrometric satellite has accurate gyros or an interferometer system, accurate astrometric observations could be obtained for a wide magnitude and wavelength range of stars and for solar system objects.

THEORIES

I include the theories of the motion of solar system objects in the fundamental reference frame, because they become an integral part of the definition of the equinox of the fundamental reference system. In the past, the theories of the motions of the solar system bodies were those of Newcomb, Brown, and Hill, augmented in part by an integration of the outer planets. The theoretical basis for the FK5 System is an accurate numerical integration, DE200/LE200, performed by the Jet Propulsion Laboratory in cooperation with the U. S. Naval Observatory. This integration has provided ephemerides of the solar system bodies which are more precise than the observational data and has provided a much more precise determination of the current reference frame. However, it presents a difficulty in formulating the definition of mean planes. The general theories, prepared at the Bureau Des Longitudes by Chapront, Bretagnon, Simon, Dvorak, and others, provide a basis for evaluating mean planes, but, the general theories are not as precise as the numerical integration and the arbitrary constants of the theories are determined from the numerical integration. In the future it would be desirable to have very accurate general theories which are valid for extended periods of time. These general theories need to be an order of magnitude more accurate than the observational data and they should be compared directly with the observations.

COMPUTATIONAL METHODS

In the past the computational methods were based on formulations designed prior to the existence of computers. Thus, it was assumed that

the observations. The time for the observations is defined by the fundamental reference system. The conversion of past observations to the present system is limited by the methods of observing, specifically by the reference catalogs used, the proper motion corrections, the day-night corrections, instrumental effects, such as flexure, and the effect of the atmosphere. The future offers the promise of milliarcsecond accuracy, particularly from space-based instruments such as Space Telescope and Hipparcos. In addition to the instrumental limitations, the ground-based observations will also be limited by the physical effects of refraction and weather.

DEFINITIONS

In the past, many of our definitions were imprecise. In fact, in many cases the adopted values were the definitions. For example, the equinox and obliquity were defined as being the values from Newcomb's Theory of the Sun. At the present time, definitions have been improved in many cases, but still mean ecliptic is imprecisely defined. Thus, the equinox and obliquity are uncertain in definition. It is hoped that in the future rigorous definitions of the ideal fundamental reference system will provide a conceptual basis for a definition to which the operational values are a satisfactory approximation.

TIME SCALES

In the past the fundamental time scale was mean solar time. This assumed that the Earth's rotation was uniform. The proof of the invalidity of that assumption lead to the introduction of Ephemeris Time. The definition of Ephemeris Time was based on Newcomb's Theory of the Sun, evaluated from Brown's Theory of the Moon and subsequently determined by atomic time. Markowitz, (1985) has determined that the second of Ephemeris Time is independent of whether the lunar or solar theories were used for its evaluation. The definition of Ephemeris Time, specifically in terms of Newcomb's theory and without relativity, led to the introduction of a new definition for a dynamical time scale.

Today International Atomic Time (TAI) provides a time reference. It is the basis for Universal Time Coordinated (UTC) which is the transmitted time and maintained in close agreement with the time of the rotation of the Earth (UT1). Terrestrial Dynamical Time (TDT) is determined from TAI by an equation. The relationship between Baricentric Dynamical Time (TDB) and (TDT) depends on the metric of the theory of relativity being utilized. This arrangement provides an independent time argument (TAI) as a common basis for the various time systems being used. It provides the freedom for an ideally defined time scale such as TDT which could be related to TAI in a more complex manner, if future observational evidence indicates a deviation from the present relationship.

The definition of universal time in terms of sidereal time is directly dependent on the equinox of the fundamental reference system.

This characteristic is a potential problem for future revision, as indicated by Williams and Melbourne, (1982). This potential problem arises because the determinations of the Earth rotation (UT1) are made by optical observations, which are tied to the fundamental reference system, radio sources which are believed to be in an inertial reference system, and solar system observations which are also thought to be in an inertial system. Therefore, subsequent corrections to the equinox, particularly time dependent corrections to the equinox (equinox motion), will lead to inconsistent corrections to the definition of sidereal time with respect to universal time.

It is not clear to me what the future time scale will be. We have established definitions of ideal time scales and we have operational time scales which approach these ideals. The future could reveal inaccuracies in our ideal definitions which would require reformulating the definitions of the ideal time scales. More likely we will learn that the approximations are not adequate and improved approximations to the ideal will be required.

RELATIVITY

In the past, relativity was only considered for planetary motions. It was incorporated as a correction in the motion of the perihelion of the planets. In the present time, relativity is introduced in the definitions of the time scales. The planetary motions are calculated based on relativistic equations. Gravitational light bending is introduced in the reductions of the optical observations and the full extent of the known relativistic effects are introduced in the reduction of the laser, radar and VLBI observations.

The future will probably require including more extensive relativistic effects as increased accuracies are achieved. There are current indications that increases in accuracies of the masses of the bodies will require a relativistic term in the length unit.

RESULTING CHARACTERISTICS

The above aspects of the fundamental reference system resulted in a system in the past that had inherent accuracy limitations. There were systematic deviations as a function of declination due to the distribution of observations, and the equinox was based on solar observations which resulted in an imprecise location of the equinox and a motion to that equinox, or a discrepancy in the equinox position which was time dependent.

The present fundamental reference system lacks observational homogeneity over the entire sky. It lacks complete data on the stars; while it provides positions and proper motions, the parallaxes and radial velocities must be found in other catalogs. There are magnitude and wavelength separations. The FK5 is based on observations of stars with magnitudes brighter than 9th magnitude. The radio coordinate system data is based on observations which correspond with optical sources of magnitude fainter than 17th. So the optical and radio coordinate systems are inherently separated. There are inconsistencies in the value of the obliquity. The obliquity defined by the planetary ephemerides is different from the IAU value which was determined from Newcomb's value. There is no definition of an ideal system, rather we have only an operational fundamental reference system.

The definition of Greenwich Mean Sidereal Time (GMST) is based on a moving equinox, so it is subject to the value of the constant of precession and the uncertainty of the equinox. On the other hand, the planetary ephemerides and the VLBI observations are tied directly to a quasi inertial coordinate system. Therefore, we are determining the rotation of the Earth based on different types of observations, which are on two different reference coordinate systems with an assumed relationship between them. As improved values in that relationship are determined, the corrections in the different cases have to be introduced accordingly.

I expect that the characteristics of the future fundamental reference system will be strongly influenced by the availability of the FK5 Catalog, which will be the best fundamental catalog ever available. There will be special purpose coordinate systems, such as the Hubble Space Telescope Guide Star Catalog, the Hipparcos Catalog, the infrared catalog, and a radio reference system catalog. All of the catalogs will, to some level of accuracy, be on the FK5 fundamental reference system. There will also be local coordinate systems, for example, a special catalog for Halley's Comet is currently in preparation. I would hope that in the future the catalogs will include all positional and motion data, including parallaxes and radial velocities. I would expect that the future will see a defined ideal fundamental reference system for which the available fundamental reference catalog will be an approximation.

THE IDEAL FUNDAMENTAL REFERENCE SYSTEM

It is probably worthwhile to try to list the desirable properties for the ideal fundamental reference system. Having listed the properties, we can consider whether we can define the system that has these properties. I think the desired properties are as follows:

a. Inertial
b. Valid for any time
c. An observable determination of the system
d. Independent of observational wavelength
e. Independent of color
f. Not dependent on magnitude or limited by magnitude
g. Independent of the observing platform
h. Not limited by observational accuracies
i. Rigorously, theoretically defined.

There are a number of possibilities for the defining geometry of the fundamental reference system. For example, the geometry could be an orthogonal coordinate system. It could be a plane and a fiducial point, such as the the equatorial plane and the equinox of the current reference system. It could be a pair of poles which would define two planes and their intersection. It could be a fiducial point and a pole, which is equivalent to a plane and a fiducial point.

In any case, directly, or indirectly, a fiducial point becomes involved. Such a fiducial point could be the equinox, such as presently defined by the solar system. It could be one QSO, or many QSOs combined to define a fiducial point. It could be a single star, or many stars. It could be a mixed catalog of optical and radio sources which define a fiducial point.

While it is possible to provide a list of the desirable properties of the ideal fundamental reference system and to give possible geometries for the definition, including possibilities for the fiducial point, it is not obvious that there is a best choice, or in fact that there is a feasible definition of this ideal fundamental reference system. Rather I have developed a list of questions which need to be considered.

1. Can the equinox, equator and obliquity be accurately defined and determined? We have operational values, but we do not have precise definitions.

2. With the interrelationships between the Earth coordinates, the astronomical coordinates, and the timescales, is it possible or desirable to ever completely modify the present system?

3. Since no points are at rest, there is a relationship between the fundamental reference system and time. Will the fundamental reference system eventually be given in terms of a time scale derived from the atomic time scale? Are observations being made in anticipation of this possibility?

4. We almost have an operational fundamental reference system, the FK5. Is it an approximation to some ideal fundamental reference system?

5. Can we rigorously define a fundamental reference system to which we can establish an operational fundamental reference system which is an approximation?

REFERENCES

Astronomical Almanac 1984

Capitaine, N., Willians, J. G. and Seidelmann, P. K. 1985 'Clarification concerning the Definition and Determination of the Celestial Ephemeris Pole' Astronomy and Astrophysics, **146**, 381.

Eichhorn, H., 1984 'Comprehensive Star Catalogs' Bull. Inform C.D.S. No. **27**.

Eichhorn, H., 1974 'Astronomy of Star Positions', Frederick Ungar Publishing Co., New York.

Explanatory Supplement to the Astronomical Ephemeris and the American Ephemeris and Nautical Almanac.

Fricke, W., 1985 'Fundamental Catalogues, Past, Present and Future' Celestial Mechanics, in press.

Fricke, W. 1979 'Methods of Compiling a Fundamental Reference System' Colloquium on European Satellite Astrometry, Padova.

Fricke, W. 1972 'Fundamental System of Positions and Proper Motions,' Annual Review of Astronomy and Astrophysics, **10**.

Markowitz, W. M. 1985 Private communication.

Mueller, I. I. 1969 Spherical and Practical Astronomy as applied to Geodesy. Frederick Ungar Publishing Co., New York.

Newcomb, S. 1906 'A Compendium of Spherical Astronomy', Dover Publications Inc., New York, New York.

Proceedings of the Sixteenth General Assembly, Grenoble 1976, Transactions of the I.A.U., Volume **XV1B**, 1977.

Williams, J. G. and Melbourne, W. G. 1982 'Comments on the Effect of Adopting New Precession and Equinox Corrections' in High-Precision Earth Rotation and Earth-Moon Dynamics, ed. by O. Calame, D. Reidel Publishing Co., Pages 293-303.

Woolard, E. W. and Clemence, G. M. (1966) Spherical Astronomy, Academic Press, New York and London.

THE EVOLUTION OF ADOPTED VALUES FOR PRECESSION *

Jay H. Lieske
Jet Propulsion Laboratory
California Institute of Technology
4800 Oak Grove Drive
Pasadena, California 91109

ABSTRACT. A survey of values of the speed of general precession in longitude is presented from the time of Bessel to the adoption of the new value by the IAU in 1976 for use in the FK5 and the J2000 reference system. The important contributions of Walter Fricke to this area of fundamental astronomy are outlined.

1. INTRODUCTION

I first corresponded with Professor Fricke while I was still a graduate student at Yale and since that time we have had many years of productive collaboration in areas of fundamental astronomy. I regard it a privilege to speak today on an area that he has greatly influenced. As we all know, Professor Fricke is one of the well-known representatives of Heidelberg and we are honoring him today on the occasion of his 70th birthday. Philip Melanchthon (1497-1560) was Martin Luther's right hand in the Reformation and it was he who drafted the Augsburg Confession for the Protestants. I happen to be a Lutheran and hold both men in high regard. Melanchthon received his initial degree from the University of Heidelberg in 1511, so we may also consider him one of the well-known products of Heidelberg. According to Thiel (1957) in _And There Was Light_ (p. 92):

> "On the occasion of the 70th birthday of Copernicus, a Protestant astronomer sent him a refutation of the absurd notion that the Earth turns round and round in the fire of the Sun, as if it were on a spit. The astronomer was acting on the behalf of Melanchthon, Germany's most famous

* Dedicated to Walter Fricke on the occasion of his 70th birthday and of his retirement as director of the Astronomisches Rechen-Institut in Heidelberg in appreciation of the insight he has given us and in anticipation of his continued research in fundamental astronomy.

Celestial Mechanics **37** (1985) 209–238.

> humanist. Melanchthon himself would have liked this dangerous nonsense officially banned. Luther commented, 'The fool would overturn all of astronomy. But in the Holy Scriptures we read that Joshua ordered the Sun to stand still not the earth.' This selfsame argument was leveled against Copernicus for more than a century. Copernicus received it with cold arrogance: 'To attack me by twisting a passage from Scripture is the resort of one who claims judgment upon things he does not understand. Mathematics is written only for mathematicians.'"

I am a Protestant astronomer, but I do not intend to roast Walter on the spit today in celebration of his 70th birthday. Based upon the reaction of some people to the new J2000 reference system, Thiel's quotation of Copernicus regarding mathematics might also be a contemporary comment!

2. EARLY DETERMINATIONS OF PRECESSION.

2.1 Investigations by Hipparchus

In Fig. 1 I present a picture of the principal precessional effects using the notation of the paper by Lieske, Lederle, Fricke and Morando (1977) which defined the new IAU precession parameters based upon the work of Fricke that was adopted by the IAU in 1976 in defining the new J2000 reference system (Contopoulos 1977). Shown are some of the basic angles that are accumulated due to precession: the luni-solar precession (ψ_A), the planetary precession (χ_A) and the general precession in longitude ($\Lambda_A - \Pi_A$), modeled after the interpretation of Andoyer (1911), as well as the equatorial precessional angles.

The early determinations of precessional motions involved the observation that the equinox slides along the fixed ecliptic. We now would describe that as the luni-solar precessional angle and it constitutes a rotation of the earth's pole about the ecliptic pole with a period of about 26,000 years. Hipparchus first observed it as a kinematical phenomenon and verified it by noting changes in the ecliptic longitude of certain stars, notably Spica (see G. J. Toomer in in his biography of Hipparchus which is contained in Gillispie 1970, Supp. pp. 207-224). Hipparchus confirmed his suspicion of the change in longitudes of stars by employing the measurements of the longitude of Spica from the Moon as determined by Timocharis (about 294 to 283 BC) in Alexandria with measurements made by himself about 160 years later. He concluded that Spica had moved two degrees in the 160-odd years since the time of Timocharis. By employing solstice observations made by Hipparchus from about 162 to 128 BC and comparing them with those of Aristarchus of Samos in 280 BC and of Meton in 432 BC, Hipparchus was able to determine the length of the year in another confirmation. He concluded that the equinoctial points move at least 1/100 degree per year (at least 3600 arcsec per century) backward through the signs of the ecliptic.

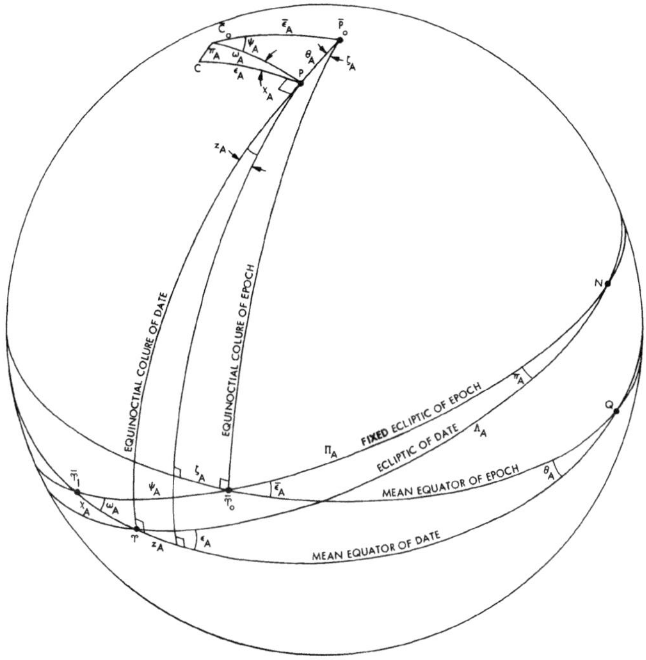

Figure 1. Illustration of principal precessional angles.

2.2 Other early determinations

Ptolemy (Grant 1852; and in G. J. Toomer's biography of Ptolemy contained in Gillispie 1970, XI, pp. 186-206), who lived about AD 100 to AD 170, reported in the Almagest that he found the same value as Hipparchus. Eighteenth century astronomers, notably Delambre, who compared the longitudes of 312 stars given by Ptolemy with those of Flamsteed about 1553 years later, suggested that Ptolemy had copied Hipparchus and merely added $2^{\circ}40'$ to Hipparchus. Delambre noted that by comparing Flamsteed to Ptolemy (an assumed interval of 1553 yr) he obtained for luni-solar precession 52.4 arcsec per year, but if Delambre assumed that Ptolemy's catalogue was really the catalogue of Hipparchus which had been adjusted by Ptolemy, then Delambre obtained (using an interval of 1820 yr) a value of 50.1 arcsec per year, but I would rather not enter into that controversy. See the paper by Ševarlić (1978) regarding old instruments and methods.

It was Copernicus who first suggested that the luni-solar precession could be described by a rotation of the earth's terrestrial pole about the ecliptic pole and the first physical description of the luni-solar precession being caused by torques exerted by the Sun and the Moon on the Earth's equatorial bulge was given by Newton (1686; Book III,

Proposition XXXIX, Problem XX, p.489ff). Newton also discovered the planetary precession, which is caused by perturbations of the planets on the Earth's orbit, causing slow changes in the orientation of the ecliptic relative to a fixed equator it and can be described as a rotation about the terrestrial pole. The first thorough solution to the problem of the precessional motions of the pole was presented by d'Alembert in 1749.

2.3 Axes of rotation

Until the 20th century the systematic effects on stellar proper motions were described by rotations about the two poles: the ecliptic pole (i.e. luni-solar precession) and the terrestrial pole (i.e. planetary precession and later including motion of a catalogue equinox). The next great step was Oort's (1927a) confirmation of Lindblad's hypothesis of galactic rotation. Oort introduced the famous "Oort constants" A and B which represent stellar motion in the galaxy. A good summary of the evolution is given by Wayman (1966). In Fig. 2 I present a diagram of the poles and spherical triangles involved in discussing rotations about the ecliptic pole (C), the terrestrial pole (T) and the galactic pole (G).

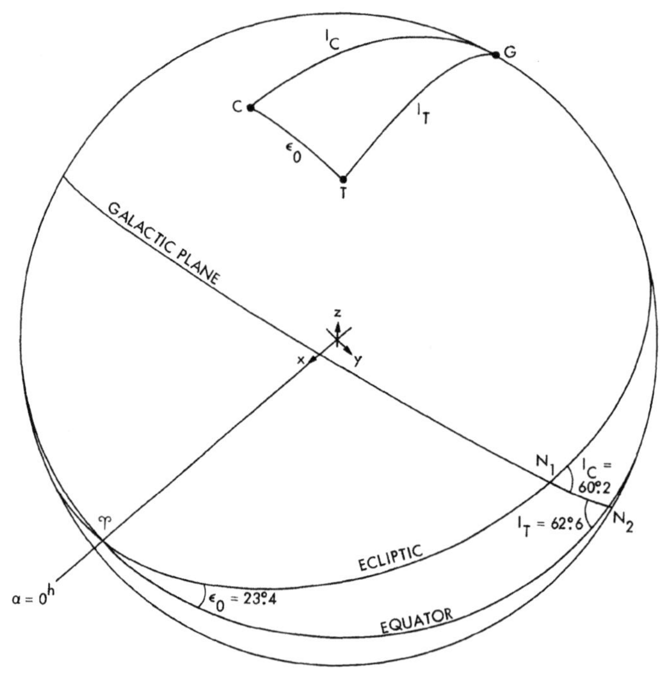

Figure 2. Celestial sphere illustrating the axes of rotation for the ecliptic pole (C), terrestrial pole (T), and galactic pole (G).

Another possible axis of rotation was suggested by Aoki (1967, 1969) and by Aoki and Kakuta (1971) who attempted to explain the then-current discrepancy between observed and theoretical values for the rate of change of obliquity. Other investigations in the geophysical aspects were given by Toomre (1966), Stacey (1973) and Rochester (1976). At the time there was some observational evidence of an unexplained change in the obliquity on the order of −0.3 arcsec/cy that was cited by numerous authors such as Spencer Jones (1932), Clemence (1943), Morgan (1950), Duncombe (1958), Duncombe and van Flandern (1976), and Fricke (1972). Duncombe and Clemence (1958) had suggested that they had found evidence in Herget's evaluation of Newcomb's Tables that such an effect existed in Newtonian mechanics.

Further study by Lieske (1970) and Laubscher (1972), based upon Fricke's encouragement and support showed that dynamically such an effect was absent from the planetary ephemerides, as does a more recent investigation by Standish (1982). The most thorough observational investigation to date was done by Wittmann (1979, 1985) who ultimately found by careful re-reduction of older data that the earlier observational result probably was fictitious in nature.

As noted by Clube (1972) and Eichhorn (1974), one of Fricke's valuable contributions to the discussion of precessional motion was Fricke's resolution of stellar motions into rotations about three axes (typically the earth equatorial xyz axes) and then after determining these motions Fricke would attempt to interpret them as due to rotations about the ecliptic pole (luni-solar precession), about the terrestrial pole (planetary precession and equinox motion) and about the galactic pole. In so doing, one could adopt various hypotheses about which poles were involved and attempt to resolve them with the observationally determined motions about the xyz axes. Earlier investigators would adopt a model ab initio and derive their results so that it would be very difficult to determine the precessional values if one slightly changed the model.

If p represents the speed of general precession, ψ the speed of luni-solar precession and χ the speed of planetary precession, then

$$p = \psi - \chi \cos \varepsilon \qquad (1)$$

where ε is the obliquity. In general, changes to the basic speeds are related by

$$\Delta p = \Delta\psi - \Delta\chi \cos \varepsilon. \qquad (2)$$

I will denote rates in tropical centuries by TC and Julian centuries by JC when it is necessary to discriminate between them.

In most cases the authors determine only the change $\Delta\psi$ in some initial value of luni-solar precession (i.e. a rotation about the ecliptic pole) and then adopt some planetary precession in order to calculate the speed of general precession in longitude. In reducing some of the results of authors who have estimated luni-solar precession, I have assumed that coupled with their value of luni-solar precession we will use the new IAU value of planetary precession in order to calculate the general precession, and I denote the value of the speed of general precession derived from the author's luni-solar precession $\Delta\psi$ and the IAU planetary precession as $\Delta p(\psi)$. If one employs the pre-1976 IAU system which was based upon Newcomb's work for the planetary precession, then the change in general precession is equal to the change in luni-solar precession minus 0.03 arcsec/cy, so that $\Delta p(\psi) = \Delta\psi - 0\overset{''}{.}03$.

3. PRECESSION DETERMINATIONS DURING THE 19th CENTURY

3.1 Possible ambiguities

Excellent summaries of precession determinations during the 19th century may be found in Dreyer (1882) and Newcomb (1895, p. 125 ff.; and 1897). When quoting values for precession errors I have always used the IAU expressions of Lieske et al. (1977) reduced to the epoch of the author and expressed in arcsec per tropical century (TC) when needed. In this way we can get some idea of how the values have changed at the epoch of the original author, but using modern values for the changes in rates. In order to obtain an equivalent value for the original authors at J2000 (using the new IAU rates to propagate the values from the author's epoch) one merely needs to add the value of Δp to the new IAU value [5029.0966 arcsec per Julian century (JC) at J2000].

At this point it may be worthwhile to bring up a question: What do we mean by a precessional rate? In current usage (Andoyer, 1911; de Sitter and Brouwer, 1938; Clemence; 1948 Lieske et al.; 1977) precessional rates are generally defined as the derivative at some time of the accumulated precessional angle measured with respect to the mean equinox of date. See the discussion in Lieske et al.(1977) regarding this point, as well as Clemence's (1948) development and his acknowledgement (1950, p. 190) of an error which was pointed out to him by Fricke's predecessor at the Rechen-Institut, A. Kopff.

In the current notation of Lieske et al. (1977), if β_A represents the amount of some angle which is accumulated due to precession, then the <u>speed</u> representing the rate of change of the angle is defined as $\beta = d\beta_A/dt\ [t = 0;\ T = T]$. It is seen (since t=0) that the speed is evaluated at the mean epoch of date and is therefore the motion with respect to the mean equinox of date. To give a concrete example of how confusion may have entered our field, I here list the values of Bessel 2 (1830) as quoted by Newcomb (1897, p. 7):

"Gen. Prec.: $50\overset{''}{.}22350 + \overset{''}{.}02243\ T$
Lunisolar: $50\overset{''}{.}36354 - \overset{''}{.}02436\ T$
T is here counted from 1800.0 in terms

of the century as the unit."

If we express these values in terms of centennial rates rather than annual rates we find that Newcomb ascribes to Bessel 2 the values

General Precession	$5022''.350 + 2''.243\ T(1800)$
Lunisolar Precession	$5036''.354 - 2''.436\ T,$

where T is measured in centuries from 1800. Simply examining these expressions one might readily be led to believe that the rate of change of general precession is 2.243 arcsec/cy^2 while the rate of change of luni-solar precession is −2.436 arcsec/cy^2. This assumption would be substantiated if one consults Newcomb's Compendium (1906), where on page 242 of the Dover edition one finds

Epoch	$d\psi/dT$	Δ
1750	$5038''.97$	
		$-1''.06$
1800	$5037''.91$	
		$-1''.07$
1850	$5036''.84$	
		$-1''.06$
1900	$5035''.78$	
		$-1''.07$
1950	$5034''.71$	
		$-1''.08$
2000	$5033''.63$	
		$-1''.08$
2050	$5032''.55$	

Newcomb uses this table to define

$$d\psi/dt = 5036''.84 - 2''.130\ T - ''.010\ T^2$$

and then states that the definitive expression for the total precession of the equinox upon the fixed ecliptic of 1850 is

$$\psi = 5036''.84\ T - 1''.065\ T^2 - ''.003\ T^2.$$

Note that Newcomb uses the origin of the <u>fixed</u> ecliptic and that ψ is the accumulated angle due to luni-solar precession from 1850. This would correspond to our ψ_A. Newcomb's $d\psi/dt$ is the modern $d\psi_A/dt$ [T=fixed]--i.e. the rate at time t with respect to the equinox at time T.

In the earlier days astronomers would generally employ the data in the manner which Newcomb presented it above: they would have two observations of right ascension at two different epochs and would divide the difference by the interval between epochs in order to obtain an average "displacement-derivative," which is measured relative to some fixed equinox. The quantity above labeled d ψ /dt by Newcomb would thus represent the displacement-derivative--that is, the rate of change of the precessional displacement ψ_A due to luni-solar precession as measured from a fixed equinox. Newcomb's ψ is the modern ψ_A [viz. the accumulated displacement] but the value ascribed by Newcomb to Bessel for luni-solar precession rate [i.e. -2.436 T] corresponds to the modern definition of d ψ_A/dt [T=0; t=T].

In modern notation Newcomb's expressions for the displacement-derivative would be written as

$$\left.\frac{d\psi_A}{dt}\right|_{\substack{T=1850\\ t=\text{time from } 1850}} = (\psi_1 + \psi_2 T + \psi_3 T^2) + 2(\psi_1' + \psi_2' T)\,t + 3\psi_1'' t^2 \qquad (3)$$

$$= \hat{\psi}_1 + \hat{\psi}_2 t + \hat{\psi}_3 t^2$$

where

$$\hat{\psi}_1 = \psi_1 + \psi_2 T + \psi_3 T^2$$
$$\hat{\psi}_2 = 2(\psi_1' + \psi_2' T)$$
$$\hat{\psi}_3 = 3\psi_1''$$

relative to the equinox of 1850, whereas the modern usage of speed would be defined as

$$\psi = \left.\frac{d\psi_A}{dt}\right|_{\substack{t=0\\ T=T}} = \psi_1 + \psi_2 T + \psi_3 T^2 \qquad (4)$$

at epoch T.

We would now say that the average "speed" at a mean epoch with respect to the equinox of the mean epoch (ignoring quadratic terms) is

$$\frac{1}{2}\left[\left.\frac{d\psi_A}{dt}\right|_{\substack{t=0\\ T=T_1}} + \left.\frac{d\psi_A}{dt}\right|_{\substack{t=0\\ T=T_2}}\right] = (\psi_1 + \psi_2)\left(\frac{T_1 + T_2}{2}\right) \qquad (5)$$

whereas in the earlier days they would say that the average "displacement-rate" at the mean epoch with respect to the fixed origin is

$$\frac{1}{2}\left[\left.\frac{d\psi_A}{dt}\right|_{\substack{t=T_1\\ T=0}} + \left.\frac{d\psi_A}{dt}\right|_{\substack{t=T_2\\ T=0}}\right] = (\psi_1 + 2\psi_2')\left(\frac{T_1 + T_2}{2}\right) \qquad (6)$$

So there may be a difficulty in interpreting the luni-solar precessional values of the 19th century. No real problem occurs with the general precession since the speed is dp_A/dt [t=0] and its "displacement-derivative" is dp_A/dt [T= 0] while the two are related by $p_2 = 2\,p_1'$. However, in the case of the luni-solar precession, the rate ψ_2 (relative to the equinox of date) changes by +.493 arcsec/cy^2 whereas the displacement-derivative changes by 2 $\psi_1' = 2 \times -1.07 = -2.14$ arcsec/cy^2

relative to a fixed equinox. Thus, if the displacement-derivatives are employed to reduce precessional accumulations, one would not find the current definition of the speed of luni-solar precession in this manner. I have estimated the differences between the "speed" and the "displacement-derivative" and find that

$$D = \langle \text{speed} \rangle - \langle \text{displacement derivative} \rangle = [\beta_2 - 2\beta_1'] (T_1 + T_2)/2 \tag{7}$$

so that in the case of general precession the result is

$$D = p_2 - 2p_1' = 2''.22 - 2 \times 1''.11 = 0 \tag{8}$$

while that for luni-solar precession is

$$D = (\psi_2 - 2\psi_1')\left(\frac{T_1 + T_2}{2}\right) = [0.49 - 2(-1.07)]\left(\frac{T_1 + T_2}{2}\right) = 2.64\left(\frac{T_1 + T_2}{2}\right) arcsec/cy \tag{9}$$

Whether or not this matters in interpreting the older values might be the subject of future study. For the present investigation I have taken the rate of general precession for 19th century authors as quoted by Newcomb (1897, pp. 1-11) and have calculated the resultant precessional values using both the values quoted by Newcomb and also using the factor given above by employing the author's mean epoch. Table 1 displays a summary of the 19th century determinations which will be discussed and Fig. 3 represents the results in graphical form.

3.2 Bessel's determinations of precession

F. W. Bessel (1818) made a lasting contribution to astronomy when he reduced the observations made by James Bradley during the years 1750-1762 in his "Fundamenta Astronomiae." Bradley (1693-1762), who succeeded Edmund Halley as Astronomer Royal, made many superb observations during his tenure in Greenwich, but never reduced them. Bessel employed his own reduction of the Bradley stars at epoch 1755 and compared the positions with those in Piazzi's catalogue of 1800. See Ševarlić et al. (1978) for a short description of older catalogues. Bessel employed 2278 stars in the RA comparison and 2429 in declination as the stars common to Bradley/Bessel and Piazzi. The time interval between catalogues was 45 years and Bessel(1815) initially determined a value (Bessel 1) for the speed of general precession for mean epoch 1777.5 which is $\Delta p = -5.71$ arcsec/cy different from the new IAU value.

Later Bessel (1830) in his "Tabulae Regiomontanae" returned to the problem of precession and revised his initial estimate. He then employed the same catalogues as earlier but supplemented the Piazzi catalogue with observations from his own catalogues made in Königsberg in 1815 and 1825, as well as introducing a newer value of nutation due to Lindenau. These changes corrected Bradley's data by $-''.5$ and Piazzi's right ascensions by $+1''.6$ and resulted in the value of general precession that is generally associated with Bessel's name. The result, when compared with the modern IAU value is $\Delta p = -2''.20$ for Bessel 2.

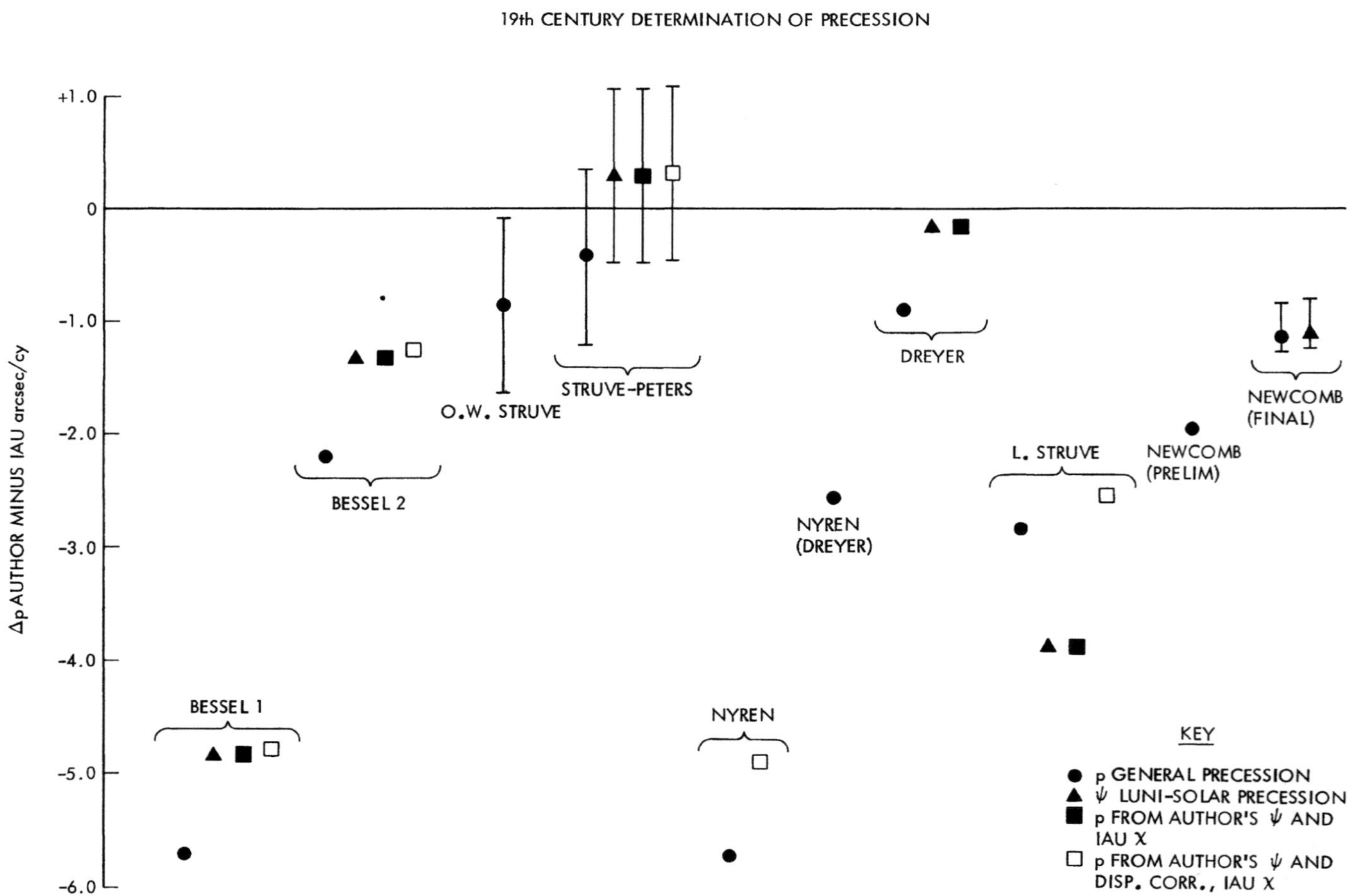

Figure 1. Nineteenth century determinations of precession

Table I. Determinations of precession in the 19th century

Value	Parameter	Author	Year	Description
$-5''.71$	Δp	Bessel 1	1815	Bessel-Bradley (1755)/Piazzi (1800)
-4.85	$\Delta\psi$			
-4.85	$\Delta p(\psi)$			
-4.78	$\Delta\hat{p}(\psi)$			mean epoch ⟨1777.5⟩
-2.20	Δp	Bessel 2	1830	Same, plus Königsberg
-1.33	$\Delta\psi$			
-1.33	$\Delta p(\psi)$			
-1.26	$\Delta\hat{p}(\psi)$			mean epoch ⟨1777.5⟩
$-0.87 \pm''.77$	Δp	O. W. Struve	1842	Bessel-Bradley (1755)/Dorpat (1825) ⟨1790⟩
$-0.44 \pm .77$	Δp	Struve-Peters	1842	Same, plus Peters' nutation
$+0.30 \pm .77$	$\Delta\psi$			
$+0.30 \pm .77$	$\Delta p(\psi)$			
$+0.31 \pm .77$	$\Delta\hat{p}(\psi)$			mean epoch ⟨1790⟩
-5.72	Δp	Nyrén	1869	Weisse (1826.7)/Schjellerup (1862.7)
-4.88	$\Delta\hat{p}(\psi)$			mean epoch ⟨1844⟩
-2.54	Δp			via Dreyer (1882)
-0.90	Δp	Dreyer	1882	Lalande (1796.7)/Schjellerup (1862.7)
-0.16	$\Delta\psi$			
-0.16	$\Delta p(\psi)$			
-2.83	Δp	L. Struve	1887	Auwers-Bradley (1755)/ Pulkovo (1855)
-3.86	$\Delta\psi$			
-3.86	$\Delta p(\psi)$			
-2.55	$\Delta\hat{p}(\psi)$			mean epoch ⟨1805⟩
$-1.95 \pm .92$	Δp	Newcomb (prel)	1895	From discussion of L. Struve, Dreyer, Nyrén
$-1''.13\ ^{+''.30}_{-.13}$	Δp	Newcomb	1897	Auwers-Bradley (1755)/ Greenwich (1862)
$-1.10\ ^{+.30}_{-.13}$	$\Delta\psi$			

For the case of Bessel 1 I find that from the printed values of general precession as quoted by Newcomb at epoch 1800, $\Delta p = -5''.71$ at 1800 [$p=5018''.83$ TC vs the IAU's $p=5024''.545$ TC]. From those of luni-solar precession I find $\Delta\psi = -4''.85$ at 1800 [$\psi = 5032''.83$ TC vs IAU $\psi = 5037''.685$ TC]. If we use this latter value along with our modern planetary precession in order to calculate the general precession, I obtain for Bessel 1 $\Delta p(\psi) = -4''.85$ at 1800 [$p=5019''.69$ TC]. If on the other hand, one employs the correction for the displacement-derivative outlined above and calculates the Bessel values at his mean epoch (1777.5) I obtain

$\Delta p = -5''.76$ [$p = 5018''.28$ TC]
$\Delta\psi = -4''.78$ [$\psi = 5032''.79$ TC]
$\Delta p(\psi) = -4''.78$ [$p = 5019''.26$ TC].

For Bessel 2 the results at 1800 from Newcomb's quotation are

$\Delta p = -2''.20$ [$p = 5022''.35$ TC]
$\Delta\psi = -1''.33$ [$\psi = 5036''.35$ TC]
$\Delta p(\psi) = -1''.33$ [$p = 5023''.21$ TC]

while the displacement-derivative correction would yield at 1777.5

$\Delta p = -2''.24$ [$p = 5021''.80$ TC]
$\Delta\psi = -1''.26$ [$\psi = 5036''.31$ TC]
$\Delta p(\psi) = -1''.26$ [$p = 5022''.78$ TC].

3.3 The Struve-Peters determination

After Bessel's determination of precession, the next major one is that generally associated with the name of Otto Wilhelm Struve (1842, 1843)--and often called the Struve-Peters precession. In 1842 Struve published a value for general precession for epoch 1790 of $\Delta p = -''.87 \pm ''.77$ [$p = 5023''.45$ TC]. This result was based upon Struve's analysis of 400 stars (75 per cent of which were double stars) observed in Dorpat at mean epoch 1825 with those same stars in Bessel's reduction of the Bradley observations at epoch 1755. It thus depends upon 70 years of data. Struve was the first to attempt to account for the motion of the Sun through space. His stars were of magnitude 1-7 and he adopted 6 mag as a norm and produced tables which gave distance as function of apparent magnitude.

The most thorough exposition of O.W. Struve's results was given by C.A.F. Peters (1842), who modified Struve's value by introducing his own value for nutation. Using Peters' values as quoted by Newcomb (1897, p. 7) we find at 1800

$\Delta p = -''.44$ [$p = 5024''.11$ TC]
$\Delta\psi = +''.30$ [$\psi = 5037''.98$ TC]
$\Delta p(\psi) = +''.30$ [$p = 5024''.84$ TC]

and if we use the displacement-derivative correction factors (1755 to 1825, mean epoch 1790) we find

$\Delta p = -''.44$ [$p = 5023''.88$ TC]
$\Delta\psi = +''.30$ [$\psi = 5037''.94$ TC]
$\Delta p(\psi) = +''.31$ [$p = 5024''.63$ TC].

The Struve-Peters precession values were in widespread use throughout the last half of the 19th century and are quite close to the modern IAU value. The difference between Peters' quotation of Struve's result is somewhat mysterious since they differ by $''.43$ per century.

3.4 The investigations of Dreyer, Nyrén and L. Struve

According to Dreyer (1882) there were "only three" determinations of precession during the 19th century that one should consider. They were those of Bessel, Struve-Peters and then Nyrén (1869). Dreyer's comment was based upon the employment of different data and methods. The results of Nyrén have generally been lost in history since his determination was drastically different from the two preceding determinations. Nyrén used 5300 faint stars within the declination band of ± 15 deg. He employed Weisse's catalog of Bessel's zones for epoch 1825 supplemented with Bessel's observations in Königsberg as well as data from Dorpat. His initial mean epoch was thus 1826.7. Nyrén then employed Schjellerup's catalogue at mean epoch 1862.7 to compare the observations. He thus determined his value over a span of 36 years. Nyrén's result is noteworthy in that he employed faint stars of mag 7-9 and that he did not compute any proper motions. He generally assumed that the stars would be distant and that their intrinsic proper motions would average out. His result, as quoted by Dreyer, on the modern IAU system at epoch 1800 is

$$\Delta p = -5''.72 \qquad [p = 5018''.82 \text{ TC} \qquad].$$

If one employs Dreyer's description of Nyrén's work and applies the displacement-derivative corrections I find at epoch 1844,

$$\Delta\psi = -4''.88 \qquad [\psi = 5033''.02 \text{ TC} \qquad]$$
$$\Delta p(\psi) = -4''.88. \qquad [p = 5020''.64 \text{ TC} \qquad].$$

This result caused much consternation when it was published. Dreyer was able to somewhat alleviate the discrepancy by reducing Nyrén's data on the Weisse and Schjellerup catalogues to Newcomb's (1872) equinox N1. Dreyer's re-evaluation of Nyrén produces at 1800,

$$\Delta p = -2''.54 \qquad [p = 5022''.0 \text{ TC} \qquad]$$

and considerably improved the situation.

Dreyer (1882) then tried his own hand at determining the precession. He employed 3300 stars within a ± 15 degree band of declination. He used Schjellerup's catalogue (mean epoch 1862.7) as did Nyrén, but Dreyer employed Lalande's observations (mean epoch 1796.7) for the earlier epoch. His span thus covered 66 years and Dreyer's results at epoch 1800 were

$$\Delta p = -''.90 \qquad [p = 5023''.65 \text{ TC} \qquad]$$
$$\Delta\psi = -''.16 \qquad [\psi = 5037''.52 \text{ TC} \qquad]$$
$$\Delta p(\psi) = -''.16 \qquad [p = 5024''.38 \text{ TC} \qquad].$$

After Dreyer's work, the next effort was made by O. W. Struve's son [Gustav Wilhelm] Ludwig Struve (1887). L. Struve used Auwers' (1888) reduction of Bradley's observations for 1755 in preliminary form. Auwers' work was the logical successor to Bessel's famous reduction of those data in "Fundamenta Astronomiae." For his second epoch, L. Struve employed catalogues made at Pulkovo 1845, 1855 and 1865. L. Struve's basic data thus spanned one century and was anticipated to be a great improvement on the earlier value of Bessel. Struve's results were (based on Newcomb's description at epoch 1850)

$\Delta p = -2''.83$	$[p = 5022''.83$ TC $]$
$\Delta\psi = -3''.86$	$[\psi = 5034''.07$ TC $]$
$\Delta p(\psi) = -3''.86$	$[p = 5021''.79$ TC $]$

while the displacement-derivative corrections would yield at epoch 1805

$\Delta p = -2''.78$	$[p = 5021''.88$ TC $]$
$\Delta\psi = -2''.55$	$[\psi = 5035''.16$ TC $]$
$\Delta p(\psi) = -2''.55$	$[p = 5022''.11$ TC $]$.

These results are not greatly different from Bessel's determination.

3.5 Simon Newcomb's researches

Finally we have arrived at the age of Newcomb. In his <u>Elements of the Four Inner Planets</u> (Newcomb, 1895), Newcomb discussed the then-available results concerning precession in order to determine a value (based upon a correction to the Struve-Peters result) which he would use in his theories for computing ephemerides of the inner planets. He primarily based his discussion on the results of L. Struve, Dreyer and Nyrén and adopted a value known as Newcomb's provisional precession. In today's terminology the results were

$$\Delta p = -1''.95 \qquad [\ p = 5023''.71 \text{ TC vs IAU } p = 5025''.656 \text{ TC }]$$

at epoch 1850. This is the value which is imbedded in Newcomb's planetary theories, even though he later adopted a different value which we today know as "Newcomb's precession." If one uses the weights as given by Newcomb in "Elements" on pp. 126-127 we can get an estimate of the uncertainty which might be attached to the above result. It turns out to be $\Delta p = -1''.95 \pm 0''.92$ approximately.

Armed with the results of his research on precession, the equinox, planetary elements and masses, and star catalogues, Newcomb journeyed to Paris in 1896 for a meeting of the heads of the national almanac offices. Since Newcomb was involved in all of the areas of fundamental astronomy at that time and since he had done research in all of them, he was the focus of turn-of-the-century fundamental astronomy and it is thus not too surprising that he more-or-less received carte blanche to produce new ephemerides, a new precession and a new star catalogue.

What we now call "Newcomb's precession" [viz. p = 5025$''$.64 per tropical century at 1900 vs the new IAU value of 5026$''$.767 TC] was developed in Newcomb's (1897) classic paper on precession. He used the Auwers/Bradley stars (L. Struve had used a preliminary version) and the Greenwich catalogues of 1860 and 1864. There were about 3000 stars in common and these formed the basis of Newcomb's analysis. Although Newcomb was aware of the possibility of galactic rotation affecting proper motions, he elected to neglect it, feeling that kinematically it was not yet well-enough determined. Newcomb's final result in today's terminology is at epoch 1900

$$\Delta p = -1''.13 \qquad [p = 5025''.64 \text{ TC vs IAU } 5026''.767 \text{ TC}]$$
$$\Delta \psi = -1''.10 \qquad [\psi = 5037''.08 \text{ TC vs IAU } 5038''.178 \text{ TC}]$$

and that precessional value was internationally adopted until the IAU in 1976 selected its new value based upon Fricke's work. Although Newcomb refrained from giving any uncertainty in his estimate of the general precession, we can obtain some value from his remarks on pp. 69-71 (Newcomb, 1897) based upon his result from right ascensions and declinations. We find that the result is

$$\Delta p = -1''.13 \;{}^{+''.30}_{-''.13}$$

or

$$-1''.26 \leq \Delta p \leq -''.83 \;.$$

4. PRECESSION DETERMINATIONS DURING THE 20th CENTURY

4.1 The era of the scientific team and committee

In the 19th century individuals, often working alone, would develop their methods and communicate them to other astronomers around the world. Often a scientist would revise his best estimates of various astronomical parameters as he progressed in his work and thus he would frequently change values as he proceeded. A case in point is LeVerrier's theories of motion of the inner planets, where three different masses of Venus are employed in the theories of the Sun, Mercury and of Mars. So it was quite difficult to determine small errors and effects of changes in such an environment when one really did not know the basic values of the astronomical parameters.

In the case of the mass of Venus, for example, it is extremely important to know its value since it contributes about 150 % of the total value of planetary precession [the other planets, notably Jupiter, generally act in the opposite direction] (Newcomb 1897, Lieske 1967, Laubscher 1972, Lieske et al. 1976). It was in this somewhat chaotic environment that the directors of the nautical almanac offices responded to Newcomb's suggestion at the Paris Conference in 1896 that the astronomers around the world adopt a common set of ephemerides and constants. This still was the era of the individual and since Newcomb had played major roles in all of the areas of fundamental astronomy, he was essentially given a free hand to come up with the definitive system. It is somewhat ironic that in the set of ephemerides and astronomical constants (and in particular that of precession), Newcomb himself used different values in his development of the theories of the inner planets (where he employed his provisional value of precession) and that he adopted a different value in his classic treatise on precession.

However, we had definitely entered the era of Team Science. It is interesting to compare the simple language employed when Commission 4 at the first IAU meeting in Rome in 1922 (Fowler, 1922) made a recommendation with the current trend of employing 'Whereas'es and 'Wherefore's in today's science. At that first meeting (p. 159) of Commission 4 [which had 8 members: H. Andoyer, E. W. Brown, L. Carnera, P. H. Cowell, W. S. Eichelberger (President), J. D. Rodrigues, G. Juvet and M. Aguilar] the recommendation was "That it would be advantageous if the directors of the National Ephemerides could arrange to adopt a uniform system of Standard Star Places."

At the second IAU meeting held in Cambridge in 1925 Eichelberger was still president of Commission 4 and he recommended that his catalogue of stars be used as the fundamental system (Fowler 1925). However, it was only at the 5th IAU meeting held in Paris in 1935 (Stratton 1935, p. 282) that Commission 4, following the recommendation proposed by Schlesinger, finally unanimously adopted a fundamental star catalogue as the standard: "It is recommended that the FK3 be the fundamental catalogue to be used for the apparent places of stars in all astronomical ephemerides."

In more recent times, Professor Fricke (1974a, 1977b, 1981) has been the guiding spirit behind the IAU's final adoption in 1976 (Contopoulos 1977; Seidelmann 1977) of a new value for precession and the new J2000 reference system. He has been the dominant figure in guiding us through many of the intricacies of developing the new system and we owe him a great deal of gratitude.

Even if the new value of precession, for example, turns out to be in error by a significant amount, we nevertheless will have gained a great deal from the experience of developing the J2000 system and finally shedding our old clothes from the Newcomb era where the "Simon Says" factor applied to everything. We had Newcomb's theories of the inner planets and the concept of dynamical time was imbedded in Newcomb's theory. Now we have independent atomic clocks that can be employed to give us the relationship between sidereal time or its cousin UT and proper time and we can determine coordinate time without resulting to an often circular set of guidelines that ultimately depended upon Newcomb's interpretations of astronomical phenomena. We had Newcomb's precession that always was subject to frustrating interpretation when one employed Newcomb's Tables of the Sun (which used a different value of "Newcomb" precession) and required an Ephemeris Time that again involved Newcomb. I recall that Myles Standish, I believe, in informal conversations would describe these interrelations as "incestuous" since they ultimately would re-focus upon Newcomb's results and interpretations. It was difficult to break out of the womb, and perhaps some mistakes have been made. Nevertheless, what we have done will be of great value to future astronomers and we owe a great deal of gratitude to Professor Fricke.

Soon after Newcomb's final value was produced there were already indications that it was seriously in error. L. Boss (1910), for example, already in 1910 had suggested a correction to Newcomb's luni-solar precession of $\Delta\psi = +.85 \pm .20$, based upon Boss' analysis of PGC proper motions.

In what follows, I will present the changes in precession found by various authors relative to the new IAU value of the speed of general precession. Newcomb's final result for general precession of $p = 5025''.64$ per tropical century at 1900 then will be given as $\Delta p = -1''.13$ since the new IAU value at 1900 is $p = 5026''.767$ per tropical century (Lieske et al. 1977). For reference, Newcomb's values at 1900 are

$$p = 5025''.64 \text{ per tropical century} \qquad [\Delta p = -1''.13]$$
$$\psi = 5037''.08 \qquad [\Delta\psi = -1''.10]$$
$$\chi = 12''.47 \qquad [\Delta\chi = +''.03]$$

while the new IAU values (at 1900 and in rates per tropical century) are

$$p = 5026''.767$$
$$\psi = 5038''.178$$
$$\chi = 12''.439$$

with

$$p = \psi - \chi \cos \varepsilon.$$

Most determinations involved corrections to the luni-solar precession. I have reduced these to corrections to general precession by adopting the author's correction to luni-solar precession and calculating the value of $\Delta p(\psi)$ by assuming the original author would accept the new value of planetary precession, so that $\Delta p(\psi) = \Delta\psi - 0''.03$.

Valuable discussions of 20th century precession determinations may be found in Wilson and Raymond (1938), Morgan (1950), Kulikov (1956), Böhme and Fricke (1965), and Fricke (1971a, 1977a), as well as in Wayman (1966) and in Fricke's extended version of his Brouwer Award Lecture (1985). Discussion of alternative formulations may be found in Murray (1978, 1983).

I will try to organize the discussion of 20th century determinations in groupings that reflect common hypotheses or common types of data. Generally all corrections were to the luni-solar precession, $\Delta\psi$, but here I will give them in terms of the speed of general precession, Δp, using the new value for planetary precession [the value is thus $\Delta p(\psi) = \Delta\psi - ''.03$ relative to the author's correction to luni-solar precession] under the old system. Table 2 and Fig. 4 contain 20th century determinations of precession.

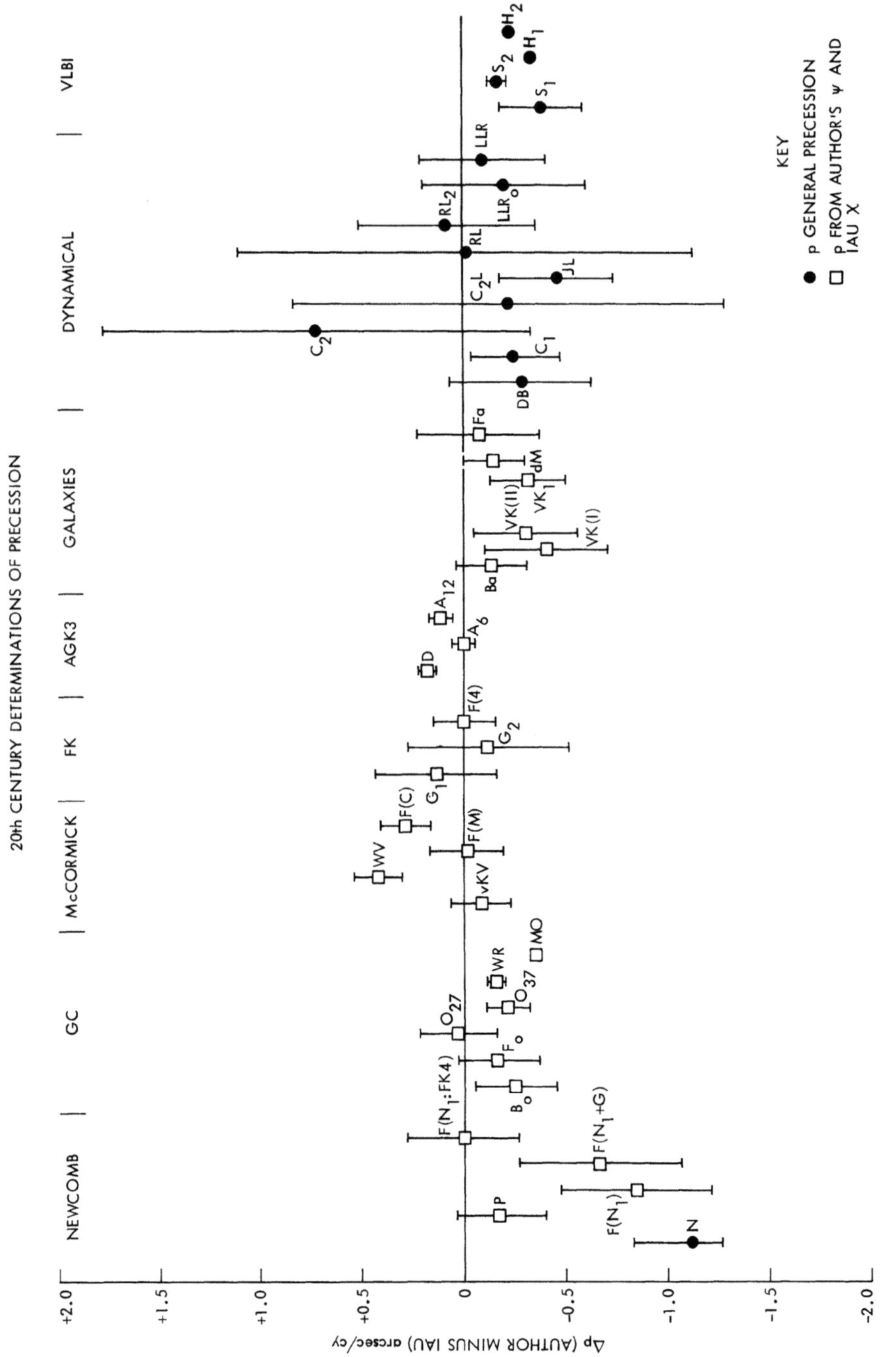

Figure 4. Twentieth century determinations of precession

Table II. Determinations of precession in the 20th century

Value	Parameter	Symbol	Author	Description
$-1''.13 \; {}^{+''.30}_{-.13}$	Δp	N	Newcomb (1897)	
$-0''.18 \pm ''.22$	$\Delta\psi$	P	Pariisky et al. (1935)	Revised weights, galactic rot.
$-0.85 \pm .37$	$\Delta\psi$	F(N1)	Fricke (1971b)	265 Bradley stars, Newcomb μ
$-0.67 \pm .37$	$\Delta\psi$	F(N1+G)		Same, plus galactic rotation
$0 \pm .27$	$\Delta\psi$	F(N1:FK4)		Same, but FK4 proper motion
$-0.25 \pm .20$	$\Delta\psi$	Bo	L. Boss (1910)	5413 PGC stars
$-0.17 \pm .20$	$\Delta\psi$	Fo	Fotheringham (1926)	Same, but galactic rotation
$+0.03 \pm .19$	$\Delta\psi$	O27	Oort (1927b)	PGC proper motions
$-0.21 \pm .10$	$\Delta\psi$	O37	Oort (1937)	Revised value
-0.35	$\Delta\psi$	MO	Morgan and Oort (1951)	Recommended value
$-0.16 \pm .04$	$\Delta\psi$	WR	Wilson and Raymond (1938)	32,000 GC stars
$-0.09 \pm .15$	$\Delta\psi$	vKV	v. de Kamp, Vyssotsky (1936)	18000 PGC stars, McCormick
$+0.41 \pm .12$	$\Delta\psi$	WV	Williams, Vyssotsky (1948)	70000 stars McCormick/Cape
$-0.02 \pm .18$	$\Delta\psi$	F(M)	Fricke (1967a)	McCormick proper motions
$+0.28 \pm .12$	$\Delta\psi$	F(C)		McCormick and Cape
$+0.13 \pm .30$	$\Delta\psi$	G1	Gliese (1940)	FK3 proper motions
$-0.12 \pm .39$	$\Delta\psi$	G2		Same, but $\mu < 6$
$0 \pm .15$	$\Delta\psi$	F(4)	Fricke (1967b)	512 FK4 stars, dist.$>$ 100 pc
$+0.17 \pm .04$	$\Delta\psi$	D	Dieckvoss (1967)	163,811 AGK3 stars
$0 \pm .05$	$\Delta\psi$	A6	Asteriadis (1977)	166,179 AGK3 stars, $\mu < 6$
$+0.11 \pm .05$	$\Delta\psi$	A12		Same, but $\mu < 12$
$-0.14 \pm .18$	$\Delta\psi$	Ba	Bakulin (1955)	870 O-B, supergiants, Ceph.
$-0.40 \pm .30$	$\Delta\psi$	VK(I)	Vasilevskis, Klemola (1971a)	Lick long-exposure plates
$-0.30 \pm .25$	$\Delta\psi$	VK(II)		short-exposure plates
$-0.32 \pm .18$	$\Delta\psi$	VK1	Vasilevskis, Klemola (1971b)	final result
$-0.07 \pm .30$	$\Delta\psi$	Fa	Fatchikhin (1970)	Soviet galaxy program
$-0.28 \pm .35$	Δp	DB	Brouwer (1950)	Sun, Mercury
$-0.25 \pm .22$	Δp	C1	Clemence (1950)	Sun, Mercury
$+0.73 \pm 1.06$	Δp	C2	Clemence (1966)	Sun, Mercury, Venus
-0.22 ± 1.06	Δp	C2L	[Lieske 1968]	Revision of above
$-0.46 \pm .28$	Δp	JL	Lieske (1970)	Eros, 1893-1966
-0.01 ± 1.12	Δp	RL	Laubscher (1976a)	Rediscuss Brouwer, Clemence
$+0.08 \pm .44$	Δp	RL2	Laubscher (1976b)	Mars 1751-1969
$-0.2 \pm .4$	$\Delta\psi$	LLRo	Dickey and Williams (1982)	Lunar Laser ranging
$-0.10 \pm .30$	$\Delta\psi$	LLR	Dickey et al. (1983)	Lunar Laser ranging 1969-82
$-0.38 \pm .20$	$\Delta\psi$	S1	Sovers et al. (1984)	VLBI 1971-80
$-0.17 \pm .02$	$\Delta\psi$	S2	Sovers (1985)	VLBI, dual frequency
$-0.28 \pm .05$	$\Delta\psi$	H1	Herring et al. (1983)	VLBI
$-0.24 \pm .01$	$\Delta\psi$	H2	Herring (1985)	VLBI, 4.5 year analysis

I will first discuss attempts to reanalyze and interpret Newcomb's data and results. Then we will proceed to attempts using PGC and GC stars, followed by investigations utilizing the McCormick/Cape catalogues, fundamental catalogues such as the FK3 and FK4, galaxies, dynamical methods and lastly, lunar laser ranging and VLBI determinations. The value adopted by the IAU is that of Fricke (1967b) based upon the analysis of FK4 proper motions.

4.2 Investigations involving Newcomb's data

In the mid-thirties Pariisky, Ogorodnikov and Fessenkov (1935) in Moscow reinvestigated Newcomb's determination under many different hypotheses. Kulikov (1956) summarizes their work as well as other Soviet investigations and gives the result $\Delta\psi = -0''.18 \pm 0''.22$ allowing for different weights and including galactic rotation. So already in the mid-thirties there was evidence that Newcomb's value ($\Delta p = -1''.13$) was seriously in error based upon alternative analyses of Newcomb's own data.

Fricke (1971b) also attempted to find out where Newcomb's error arose. He did not employ Newcomb's data, but used a sub-set of the data: 265 Bradley stars with distances greater than 100 pc that appear in the FK4 and FK4 Sup and that were in the list of 512 FK4/FK4 Sup stars that were used by Fricke (1967b) in determining his value of precession that has been adopted by the IAU. Fricke found that most of the problems of Newcomb stem from the system of proper motions due to Auwers which Newcomb employed. In his analysis Fricke obtained $\Delta\psi = -0''.85 \pm 0''.37$ using Newcomb's (1898) Fundamental Catalogue proper motions, and he found $\Delta\psi = -0''.67 \pm 0''.37$ if he included galactic rotation. He found $\Delta\psi = 0 \pm 0''.27$ if he used FK4 proper motions and allowed for galactic rotation. Although Fricke did not use the 3000 stars which Newcomb had employed, he had a controlled group and his results suggested that while a minor part of Newcomb's error was due to neglect of galactic rotation, most of it was due to deficiencies in the proper motions of the Auwers/Bradley catalogue and errors in Newcomb's equinox N1.

4.3 Investigations based on PGC and GC and McCormick/Cape catalogues

As noted earlier, Boss (1910) found a value of $\Delta\psi = -0''.25 \pm 0''.20$ based upon analysis of the proper motions of 5413 stars in the PGC. Fotheringham (1926) re-evaluated Boss' investigation and after allowing for galactic rotation found a value $\Delta\psi = -0''.17 \pm 0''.20$. Oort (1927b), who had experimentally confirmed Lindblad's hypothesis of galactic rotation and introduced the "Oort constants" found from an analysis of PGC proper motions the result $\Delta\psi = +0''.03 \pm 0''.19$. He later (1937) obtained a value $\Delta\psi = -0''.21 \pm 0''.10$ and ultimately recommended (Morgan and Oort 1951) a value of $\Delta\psi = -0''.35$. Other investigations were done by Wilson and Raymond (1938) using 32,000 GC stars who obtained $\Delta\psi = -0''.16 \pm 0''.04$.

The next important series of determinations concerns investigations which involve the McCormick and Cape photographic catalogues. Van de

Kamp and Vyssotsky (1936) found from an analysis of 18,000 faint PGC stars in the McCormick photographic catalogue that $\Delta\psi=-''.09 \pm ''.15$. Later on Williams and Vyssotsky (1948) found a fairly large value $\Delta\psi=+''.41 \pm ''.12$ from analysis of 70,000 faint stars in the McCormick and Cape catalogues. It was in an attempt to confirm the Williams and Vyssotsky result that Fricke (1967a) launched his lengthy investigations into precession. Fricke found from an analysis of McCormick proper motions that $\Delta\psi=-''.02 \pm ''.18$ while he found from McCormick and Cape the result $\Delta\psi =+''.28 \pm ''.12$ using the FK4 system. His final result was that

$$-''.02 \pm ''.18 \leq \Delta\psi \leq +''.28 \pm ''.12 .$$

4.4 Investigations involving FK3 ,FK4 and AGK3

We next proceed to investigations involving the proper motion systems of the FK3 (Kopff, 1937) and the FK4 (Fricke and Kopff et al. 1963). In an investigation in 1940 Gliese (1940) found from an analysis of FK3 proper motions the value $\Delta\psi=+''.13 \pm ''.30$ if he employed all the FK3 stars and he found the value $\Delta\psi=-''.12 \pm ''.39$ if he only employed stars with proper motion less than 6 arcsec. In the determination which has directly led to adoption by the IAU of a new speed of general precession, Fricke (1967b) analyzed the proper motions of 512 stars in the FK4 and FK4 Sup which are at distances greater than 100 pc. His result is the classical value $\Delta\psi=0 \pm ''.15$.

In order to obtain many stars which could be used in the analysis, several investigations were conducted which involve the AGK3. Dieckvoss (1967) in an analysis of the preliminary proper motions of 163,811 stars in the AGK3 divided into 42 groups by magnitude and spectral class obtained the result $\Delta\psi=+''.17 \pm ''.04$. Later, one of Fricke's colleagues Asteriadis (1977) made the most recent discussion of the AGK3 and employing 166,179 AGK3 stars dimmer than photographic magnitude 8 obtained the value $\Delta\psi=0 \pm ''.05$ if he limited the stars to those with proper motions less than 6 arcsec. Asteriadis obtained $\Delta\psi=+''.11 \pm ''.05$ if he included stars with proper motions less than 12 arcsec.

4.5 Investigations involving galaxies

Very early in the history of astronomy it was known how beneficial it would be to measure proper motions relative to distant galaxies. Clemence (1966) in his George Darwin Lecture mentioned that D. W. Dewhirst brought to Clemence's attention that Herschel (1785) recognized the importance of using galaxies as a reference. According to Clemence,

> "Herschel (concerning two condensed nebulae that he has recently discovered) said 'If...they belong most probably to the class of nebulae, they may then be expected to keep their situation better than any one of the stars belonging to our system, on account of their being probably at a very great distance. Now to have a fixed point somewhere in the heavens, to which the motions of the rest may be

> referred, is certainly of considerable consequence in astronomy; and both these bodies are bright and small enough to enter that end.'
> We now know that these objects are planetary nebulae, but Herschel's recognition of the utility of small bright galaxies as a reference frame is quite clear."

In early attempts to move away from using stars which are in our local area Bakulin (1955) employed 870 stars of Class O and B0–B5, supergiants and long-period Cepheids to obtain a value of $\Delta\psi=-''.14 \pm ''.18$.

In the United States, Wright (1950) proposed referring photographic positions of stars to galaxies and he instituted a program at Lick Observatory for this purpose. The first epoch plates were started in 1947 and were completed in 1954 (Vasilevskis 1954). At the IAU Colloquium held in Heidelberg in 1970 Vasilevskis and Klemola (1971a) reported values $\Delta\psi=-''.40 \pm ''.30$ from the long-exposure plates using fainter stars and $\Delta\psi=-''.30 \pm ''.25$ from the short-exposure plates using brighter stars. In a later paper (1971b) they reported a value of $\Delta\psi=-''.32 \pm ''.18$. The most recent work with the Lick material has been performed by du Mont (1978) at the suggestion of Fricke. Du Mont finds the value of $\Delta\psi=-''.15 \pm ''.15$, based upon the analysis of 8072 Lick stars.

Soviet efforts in use of galaxies were noted by Deutsch (1954) who explained the galactic program in the USSR. Deutsch (1982) reports that Gerassimovitch and Dneprovsky in 1933 advanced the idea of construction of new fundamental catalogues of faint stars whose proper motions were to be related to extragalactic nebulae. The Soviet first-epoch plates were completed in 1961 (Deutsch 1962) and a summary of their program is given by Deutsch (1982) in the second part of "Epitome Fundamentorum Astronomiae" by Teleki and Ševarlić (1982). Fatchikhin (1970a, b) gives the result of $\Delta\psi=-''.07 \pm ''.30$.

4.6 Investigations by the Dynamical Method

The value of precession can be deduced via the dynamical method by analyzing the motion of solar system bodies expressed in an inertial frame and comparing the predicted positions with those actually observed in an astronomical frame. The differential equations describing the planet's motion in the gravitational theory define a presumed inertial frame. By analyzing optical observations of the planets (where the observations would show a slow rotation relative to the dynamical frame) one can determine precession by the dynamical method. A good summary is given by Laubscher (1976a). The early attempts did not show much promise and Brouwer (1950) who based his discussion on an analysis of solar data (Morgan 1945) and the results of Clemence's (1943) discussion of Mercury observations found the value $\Delta p=-''.28 \pm ''.35$ to the general precession. Clemence (1950) mentioned that he had earlier obtained $\Delta p=-''.25 \pm ''.22$ from similar data.

In his George Darwin Lecture Clemence (1966) reported that he had recently completed a solution based upon his (1943) investigation of Mercury, upon Morgan's solar results and upon Duncombe's (1958) discussion of Venus, excluding transit results. At that time Clemence gave the value $\Delta p=+''.73 \pm 1''.06$ [i.e. a correction to Newcomb of $1''.86$]. This rather large discrepancy with earlier work led me in an unpublished work (since at that time I was a student of Clemence's!) about 1968 to investigate Clemence's result. I found that Clemence had included one equation in his solution with the wrong sign and when I employed the IAU masses I found $\Delta p=-''.22 \pm 1''.06$.

In an investigation involving Eros Lieske (1970) obtained the result $\Delta p=-''.46 \pm ''.28$, but he also included in the solution an estimate of the obliquity-rate error. Laubscher (1976a) investigated the precession using data from the Sun, Mercury and Venus and published the value $\Delta p=-''.01 \pm 1''.12$, although he really found that the data were not very sensitive to precession and that he could not <u>dis</u>prove Fricke's value. The most recent dynamical investigation by Laubscher (1976b) was based upon the motion of Mars from 1751-1969 and yielded the correction of $\Delta p=+''.08 \pm 0''.44$, in reasonable agreement with the newly adopted IAU value.

5. LASER RANGING AND VERY LONG BASELINE INTERFEROMETRY (VLBI)

Recently values have begun to be determined from analysis of lunar laser ranging and future results from this method look bright. In an early result Dickey and Williams (1982) reported a preliminary value of $\Delta p=-''.2 \pm ''.4$ from lunar laser ranging. Later in Dickey, and Williams et al. (1983) they produced the value $\Delta p=-''.10 \pm ''.30$, based upon lunar laser ranging data from 1969 through 1982. While they actually obtain a correction $\Delta\psi$ to the luni-solar precession, I quote their results here in terms of the general precession since they have used the J2000 system and have thus employed the revision to planetary precession.

After the preliminary results of Elsmore (1976) indicated the potential of VLBI for determinations of precession, Walter (1977) summarized the possible accuracies.

Very recently VLBI has started to produce results on precession (actually on luni-solar precession and nutation since at this time they cannot yet separate the two effects very well). Using the IAU standards Sovers et al. (1984) at JPL reported from an analysis of VLBI data from 1971 to 1980 the value $\Delta p=-''.38 \pm ''.20$, while a more recent result of Sovers (1985) using dual-frequency data is $\Delta p=-''.17 \pm ''.02$.

On the East Coast of the United States, Herring and his colleagues (1983) have reported a value of $\sin\varepsilon\,\Delta p=-''.11 \pm ''.02$ or $\Delta p=-''.28 \pm ''.05$. More recently Herring (1985) reported a value of $\sin\varepsilon\,\Delta p = -''.095 \pm ''.005$ or $\Delta p = -''.24 \pm ''.01$, although it is perhaps too early to characterize the correction as being solely due to an error in precession. Herring analyzed 4.5 years of VLBI data and, after estimating corrections to forced nutation terms with periods of a year or less, he found the secular correction given above which can also contain corrections to terms in the nutation with periods greater than a year. The total correction found by Herring, for example, could be ascribed to an error in the 18-year nutation term on the order of 0.007 arcsec, and it will remain for future investigations to separate the effects of long period nutation terms from precession.

The potential of VLBI data is very promising and it will be interesting to watch how the values converge as they become better determined. Again, I have listed these corrections in terms of general precession, Δp, even though the authors actually derive corrections to the luni-solar precession since they have adopted the J2000 system and the revised planetary precession [i.e. $\Delta p(\psi) = \Delta\psi$]. Adoption of the new J2000 system should help in this regard, since it was only after the various VLBI teams adopted a common preliminary description of the new system [Kaplan (Ed.)1981] which appears as a Supplement to the Astronomical Almanac for 1984, that the discordances among the various VLBI teams were reduced.

5. SUMMARY

We have outlined the evolution of various determinations of the speed of general precession in longitude. In the development of the new IAU system Professor Fricke has played a major role and we are indeed in his debt.

According to the biography of Newcomb written by Brian Marsden and appearing in Gillispie (1970, Vol X, p. 33), Newcomb was charged with producing the new astronomical constants that have been in widespread use during this century. However, Newcomb officially had to retire from his position on his birthday in 1897--before many of the important works which we ascribe to him were completed. In Newcomb's case arrangements were made for him to continue his work on a consulting basis and for that wisdom of the U. S. government we can be very grateful. By the same token, we hope that Walter Fricke will continue his important and far-reaching researches in fundamental astronomy long after he officially retires as director of the Astronomisches Rechen-Institut. Glück auf!

This paper represents the results of one phase of research conducted at the Jet Propulsion Laboratory/Caltech under contract with the National Aeronautics and Space Administration.

REFERENCES

Andoyer, H.: 1911, Bull. Astron. **28**, 67-76

Aoki, S.: 1967, Publ. Astron. Soc. Japan **19**, 585

Aoki, S.: 1969, Astron. J. **74**, 284

Aoki, S., Kakuta, C.: 1971, Celest. Mech. **4**, 171-181 (IAU Colloquium No. 9, Heidelberg, 1970)

Asteriadis, G.: 1977, Astron. Astrophys. **56**, 25-38

Auwers, A.: 1888, 'Neue Reduktion der Bradleyschen Beobachtungen,' Band 3, St. Petersburg

Bakulin, P. I.: 1955, Tr. Gos. Astron. Inst. Shternberg **26**

Bessel, F. W.: 1815, Mem. Berlin Acad. Sci. **I**, 262-285 in R. Engelmann, 'Bessel's Abhandlungen'

Bessel, F. W. 1818: 'Fundamenta Astronomiae pro anno 1755 deducta ex observationibus viri incomparabilis James Bradley in specola astronomiae Grenovicensi per annos 1750-1762 institutis.' Regiomonti

Bessel, F. W.: 1830, 'Tabulae Regiomontanae Reductionum Observationum Astronomicarum ab Anno 1750 usque ad Annum 1850 Computate.' Regiomonti, Petersburg

Böhme, S., Fricke, W.: 1965, Bull. Astron. **25**, 269 [IAU Symposium No. 21: The System of Astronomical Constants, J. Kovalevsky Ed., Gauthier-Villars]

Boss, L.: 1910, Astron. J. **26**, 111-122

Brouwer, D.: 1950, Bull. Astron. **15**, 165-180

Clemence, G. M.: 1943, Astron. Pap. **11**, Part 1

Clemence, G. M.: 1948, Astron. J. **53**, 169-179

Clemence, G. M.: 1950, Bull. Astron. **15**, 181

Clemence, G. M.: 1966, Q. J. R. Astron. Soc. **7**, 10-21 (George Darwin Lecture)

Clube, S. V. M.: 1972, Mon. Not. R. Astron. Soc. **159**, 289-314

Contopoulos, G. (Ed.): 1977, Trans. IAU **16B**

de Sitter, W., Brouwer, D.: 1938, Bull. Astron. Inst. Neth. **8**, 213-231

Deutsch, A. N.: 1954, Trans. IAU **8**, 789

Deutsch, A. N.: 1962, Trans. IAU **11B**, 403

Deutsch, A. N.: 1982, 'Photographic Part of the Catalogue of Faint Stars' in Epitome Fundamentorum Astronomiae, Pars II (Teleki, G. Ševarlić, B. Eds) Publ. Astron. Ops. Beogradu, No. 29 , pp. 41–52 [summary of Soviet galactic work]

Dieckvoss, W. : 1967, Astron. Nachr. **290**, 141

Dickey, J. O., Williams, J. G.: 1982, 'Correction to the 1976 IAU value of the Precession Constant' report to NASA Geodynamics Rev., Washington

Dickey, J. O., Williams, J. G., Newhall, X X, Yoder, C. F.: 1983: 'Applications of Lunar Laser Ranging' in Proc. Int. Assoc. Geod. Symp. (H. Kautzleben, J. D. Bossler, G. Lachapelle, A. M. Wassef Eds.), Ohio State Univ. 2, 509–521

Dreyer, J. L. E.: 1882, 'A New Determination of the Constant of Precession' Copernicus (Dublin) **II**, 135–155

du Mont, B.: 1978, Astron. Astrophys. **66**, 441–451

Duncombe, R. L.: 1958, Astron. Pap. **16**, Part 1

Duncombe, R. L., Clemence, G. M.: 1958, Astron. J. **63**, 456

Duncombe, R. L., van Flandern, T. C.: 1976, Astron. J. **81**, 281

Eichhorn, H.: 1974, Astronomy of Star Positions, F. Ungar, N. Y.

Elsmore, B.: 1976, Mon. Not. R. Astron. Soc. **177**, 291–298

Fatchikhin, N. V.: 1970a, Izv. Gl. Astron. Obs. Pulkovo **185**, 93

Fatchikhin, N. V.: 1970b, Astron. Zh. **47**, 619 [English transl. in Soviet Astron. **14**, 495]

Fotheringham, J. K.: 1926, Mon. Not. R. Astron. Soc. **86**, 414

Fowler, A. (Ed.): 1922, Trans. IAU **I**

Fowler, A. (Ed.): 1925, Trans. IAU **II**

Fricke, W., Kopff, A., Gliese, W., Gondolatsch, F., Lederle, T., Nowacki, H, Strobel, W., Stumpff, P.: 1963, 'Fourth Fundamental Catalogue' Veröff. Astron. Rechen-Institut, Heidelberg, Nr. 10

Fricke, W.: 1967a, Astron. J. **72**, 642–649

Fricke, W.: 1967b, Astron. J. **72**, 1368–1379

Fricke, W.: 1971a, Celest. Mech. 4, 150 (IAU Colloquium No. 9, Heidelberg, 1970)

Fricke, W.: 1971b, Astron. Astrophys. **13**, 298-308

Fricke, W.: 1972, 'Fundamental Systems of Position and Proper Motions' in Ann. Review Astron. Astrophys. **10**, 101-128

Fricke, W.: 1974a, 'Pro and Contra Changes in the Conventional Values of Precession' in Highlights of Astronomy, G. Contopoulos (Ed.)

Fricke, W.: 1977a, 'Basic Material for the Determination of Precession and of Galactic Rotation and a Review of Methods and Results' Veröff. Astron. Rechen-Institut, Heidelberg, No. 28

Fricke, W.: 1977b, Astron. Astrophys. **54**, 363-366

Fricke, W.: 1981, 'Definition and Practical Realization of the Reference Frame in the FK5--the Role of Planetary Dynamics and Stellar Kinematics in the Definition' in Reference Coordinate Systems for Earth Dynamics (E. M. Gaposchkin, B. Kolaczek, Eds.) 331-340

Fricke, W.: 1985, 'Fundamental Catalogues: Past, Present and Future' [Brouwer Award Lecture 1983] Celest. Mech. (In Press)

Gillispie, C. C. (Ed.): 1970, Dictionary of Scientific Biography, Chas. Scribner, N. Y.

Gliese, W.: 1940, Astron. Nachr. **270**, 127-144

Grant, R.: 1852, History of Physical Astronomy (Johnson Reprint, 1966)

Herring, T. A., Shapiro, I. I., Webber, J. C., Rogers, A. E. E., Corey, B. E., Whitney, A. R., Cappallo, R. J., Hinteregger, H. F., Clark, T. A., Ryan, J. W., Ma, C., Schulper, B. R., Pigg, J. C., Knight, C. A., Shaffer, D. B., Vandenberg, N. R.: 1983, 'Geodesy by Radio Interferometry: Studies of the nutations of the Earth's Rotation Axis' EOS Trans. Am. Geophys. Union **64**, 674

Herring, T. A.: 1985: Personal communication [$\sin\varepsilon\ \Delta\psi = -.095 \pm .005$]

Herschel, W.: 1785: Philos. Trans. R. Soc. **75**, 213

Kaplan, G. H. (Ed.): 1981, U. S. Naval Obs. Circ. No. 163

Kopff, A.: 1937, 'Dritter Fundamentalkatalog des Berliner Astronomischen Jahrbuchs' I Teil: Veröff. Astron. Rechen-Institut Berlin No. 54 (1937). II Teil: Abh. Preuss. Akad. Wiss. Phys.-Math. Nr. 3 (1938)

Kulikov, K. A.: 1956, Fundamental'nye postoyannye astronomii, Moscow. English translation in Fundamental Constants of Astronomy, NASA

Tech. Transl. F-143 (1964)

Laubscher, R. E.: 1972, Astron. Astrophys. **20**, 407-414

Laubscher, R. E.: 1976a, Astron. Astrophys. **51**, 9-12

Laubscher, R. E.: 1976b, Astron. Astrophys. **51**, 35

Lieske, J. H.: 1967, 'Expressions for the Precession Quantities and their Partial Derivatives' NASA Tech. Rept. 32-1044, JPL

Lieske, J. H.: 1970, Astron. Astrophys. **5**, 90-101

Lieske, J. H., Lederle, T., Fricke, W., Morando, B.: 1977, Astron. Astrophys. **58**, 1-16

Morgan, H. R.: 1945, Astron. J. **51**, 127

Morgan, H. R.: 1950, Bull. Astron. **15**, 211

Morgan, H. R., Oort, J. H.: 1951, Bull. Astron. Inst. Neth. **11**, 379-384

Murray, C. A.: 1978, Mon. Not. R. Astron. Soc. **183**, 677-685

Murray, C. A.: 1983, Vectorial Astrometry (A. Hilger, Bristol)

Newcomb, S.: 1872, 'Right Ascensions of the Equatorial Stars,' Washington Observ. for 1870, Appendix III [defines N1 equinox]

Newcomb, S.: 1895, Elements of the Four Inner Planets (Supplement to Am. Ephem. Naut. Alm. 1897), Washington

Newcomb, S.: 1897, 'The Precessional Constant' Astron. Pap. Washington Vol. **8**, Part 1

Newcomb, S.: 1898, 'Catalogue of Fundamental Stars' Astron. Pap. **8**, Part 2 [Newcomb's Fundamental Catalogue]

Newcomb, S.: 1906, Compendium of Spherical Astronomy (Macmillan 1906 Dover Reprint 1960)

Newton, I.: 1686, Principia. Revision of Motte's translation by F. Cajori: 1934, Sir Isaac Newton's Mathematical Principles, U. Calif. Press, 1960, 4th printing

Nyrén, M.: 1869, 'Försök til bestämning af Precessionskonstanten medelst ljussvaga stjärnor' Uppsala. [Also Melan. math. astron. **IV**, 590]

Oort, J. H.: 1927a, Bull. Astron. Inst. Neth. **3**, 275

Oort, J. H.: 1927b, Bull. Astron. Inst. Neth. **4**, 79

Oort, J. H.: 1937, Bull. Astron. Inst. Neth. **8**, 149

Pariisky, N. N., Ogorodnikov, F. K., Fessenkov, V. G.: 1935, Tr. Gos. Astron. Inst. Shternberg [Publ. Sternberg State Astron. Inst.] **6**, Nr. 1, 104-186 (in Russian), 188-194 (in English)

Peters, C. A. F.: 1842, 'Numerus constans nutationis ... in specula Dorpatensi annis 1832-1838 observatis deductus', p. 160, St. Petersburg

Rochester, M. G.: 1976, Geophys. J. R. Astron. Soc. **46**, 109-126

Seidelmann, P. K.: 1977, Celest. Mech. **16**, 165

Ševarlić, B. M.:1978, 'Fundamental Astrometry--A Look Through the Past' in Publ. Dept. Astron. Beograd, No. 7 , pp.7-30 (B. Ševarlić, G. Teleki Eds)

Ševarlić, B., Teleki, G.: 1978, 'Epitome Fundamentorum Astronomiae, Pars 1: Catalogues of Star Positions' Publ. Dept. Astron. Beograd, No. 7 [See Teleki and Ševarlić for Part 2]

Ševarlić, B. M., Teleki, G., Szádeczky-Kardoss, G.: 1978, 'Bibliography of the Catalogues of Star Positions' in Publ. Dept. Astron. Beograd, No. 7 (B. Ševarlić, G. Teleki Eds.) , pp. 69-272 [listing of old catalogues]

Sovers, O. J., Thomas, J. B., Fanselow, J. L., Cohen, E. J., Purcell, G. H. Jr., Rogstad, D. H., Skjerve, L. J., Spitzmesser, D. J.: 1984, J. Geophys. Res. **89**, No. B9, 7597-7607

Sovers, O. J.: 1985, Personal Communication [-.17±.02 arcsec/cy]

Spencer Jones, H.: 1932, Ann. Cape. Obs. **13**, No. 3, 1

Spencer Jones, H.: 1950, Bull. Astron. **15**, 247

Stacey, F. D.: 1973, Geophys. J. R. Astron. Soc. **33**, 47-55

Standish, E. M.: 1982, Astron. Astrophys. **114**, 297-302

Stratton, F. J. M. (Ed.): 1935, Trans. IAU V

Struve, L.: 1887, 'Bestimmung der Constante der Präcession und der eigenen Bewegung des Sonnensystems' Mem. Acad. Imp. St. Petersbourg, VII Ser. 35, No. 3 [Gustav Wilhelm Ludwig Struve]

Struve, O. W.: 1842, 'Bestimmung der Constante der Präcession mit Berücksichtigung der eigenen Bewegung des Sonnensystems' Mem. Acad. Imp. St. Petersbourg, VI Ser., 5, No. 1, 17-124

Struve, O. W.: 1843, Astron. Nachr. **21**, Nr. 485, 65

---: 1983: Supplement to Astronomical Almanac 1984, 'The Improved IAU System'

Teleki, G., Ševarlić, B.: 1982, 'Epitome Fundamentorum Astronomiae, Pars II: Photographic Catalogues and Charts of Star Positions', Publ. Astron. Ops. Beogradu, No. 29 [old catalogues. See Ševarlić and Teleki for Part I]

Thiel, R.: 1957, And There Was Light. Mentor translation by R. and C. Winston of Und Es Ward Licht

Toomre, A.: 1966, 'On the coupling of the Earth's core and mantle during the 26,000-year precession' in B. G. Marsden, A. G. W. Cameron (Eds.) The Earth-Moon System, Plenum Press, 33-45

van de Kamp, P., Vyssotsky, A. N.: 1936, Astron. J. **45**, 167

Vasilevskis, S.:1954, Astron. J. **59**, 40

Vasilevskis, S., Klemola, A. R.: 1971a, Celest. Mech. 4, 163-170 [IAU Colloquium No. 9, Heidelberg 1970]

Vasilevskis, S., Klemola, A. R.: 1971b, Astron. J. **76**, 508-512

Walter, H. G.: 1977, Astron. Astrophys. **59**, 433-440

Wayman, P. A.: 1966, Q. J. R. Astron. Soc. **7**, 138

Williams, E. T. R., Vyssotsky, A. N.: 1948, Astron. J. **53**, 63

Wilson, R. E., Raymond, H.: 1938, Astron. J. **47**, 49-68

Wittmann, A. D.: 1979, Astron. Astrophys. **73**, 129

Wittmann, A. D.: 1985, Sterne Weltraum **24**, 24

Wright, W. H.: 1950: Proc. Am. Philos. Soc. **94**, 1

ON THE ORIENTATION OF EPHEMERIS REFERENCE FRAMES

E. Myles Standish Jr.
Jet Propulsion Laboratory/Caltech
Pasadena, California 91109

ABSTRACT. One may construct complete planetary and lunar ephemerides, referred to the equator and dynamical equinox of some epoch, strictly from ranging data alone. Such an ephemeris would be completely independent from any optical data and therefore independent of any stellar catalogue. By using such an ephemeris to then analyse optical observations, one could theoretically derive many of the pertinent features of the catalogue system to which the optical observations are referred. Such features include the equinox offset, equinox motion and systematic proper motion errors. In practice, the optical observations are used in the fitting process, but essentially the same determinations may be made.

This paper presents estimates of the equinox offset and equinox motion of the FK4 as determined by the ephemeris fitting process and compares them with corresponding determinations by Fricke. No significant differences are found. Further, it is indicated how one may also estimate a value for precession and the value of the obliquity from the ephemerides. These values are also compared with the presently adopted ones.

I. INTRODUCTION

There are discussions in the literature referring to the creation of lunar and planetary ephemerides (Standish, 1982a, 1985 and Newhall et al., 1983). These describe how each of the various types of observational data determines different features of the ephemerides, such as relative positions, relative motions, orientation, absolute motions with respect to inertial space, obliquity, etc. For the inner planets, it is seen that none of these features is significantly influenced by the inclusion of optical observational data with the sole exception of the location of the zero-point of the right ascension system. In practice, the optical data serve only to orient the reference frame of the ephemerides onto the right ascension system of the stellar catalogue (FK4) upon which these optical data are based.

Celestial Mechanics **37** (1985) 239–242.

Theoretically, however, one could omit the optical data entirely and, instead, refer the right ascension system to the dynamical equinox, a location which may be derived immediately from the ephemeris itself. Such procedures have been used by Standish (1982b) and by Chapront-Touze and Chapront (1983).

If there are systematic errors in the optical observations, one may expect to see these trends appearing in the residuals computed against the ephemerides, since the motions of the bodies are not strongly influenced by the optical data. If these systematic errors could be attributed to errors in the positions and proper motions of the stellar catalogue, then it would be possible to analyse the optical residuals in order to derive corrections to the stellar catalogue.

In the creation process of the J2000-based FK5, Fricke (1971) has determined a correction to the previously adopted value of precession; he has also determined the equinox offset and equinox motion of the present standard catalogue, the FK4 (Fricke, 1982). These determinations are made using observational data which is almost entirely optical; as such, they are nearly independent from the ephemeris creation methods referred to above.

It then makes sense to compare the results of the two sets of determinations.

II. COMPARISONS

Table I presents a comparison between the various quantities determined from the ephemerides and the corresponding values determined by Fricke for precession (1971), for equinox offset and motion (1982), and the IAU value of obliquity.

A. Precession and Equinox Motion

A correction to precession produces the following systematic trends in the optical residuals:

$$\Delta\alpha = (\Delta p \cos\epsilon + \Delta p \sin\epsilon \sin\alpha \tan\delta)\ T \quad \text{and}$$
$$\Delta\delta = (\Delta p \sin\epsilon \cos\alpha)\ T,$$

where ϵ is the obliquity and Δp is the correction to the precession constant. The trend resulting from the equinox motion is of the form

$$\Delta\alpha = -\dot{E}\ T,$$

where $\dot{E}$ is the motion of the equinox.

The major part of the trend in right ascension is the secular part, resulting from the combination $(\Delta p \cos\epsilon - \dot{E})\ T$. Thus, the only sensitivity to precession alone comes from the weaker mixed-secular terms. Unfortunately, the bulk of the optical observations which may be used to determine these effects are those of the daytime objects (i.e., the Sun, Mercury and Venus). Since these observations are known to be highly subject to their own peculiar observational errors, it is not possible to determine an accurate predession correction using the ephemerides. As such, Fricke's value of 1".1/cty for the correction to precession has been adopted in the ephemeris solutions.

It may be noted here, though, that the lunar laser-ranging observations are indeed sensitive to precession. Within a few years, after the lunar observations have spanned a full 18.6-year period of the nodal regression, these data will provide an improved determination of this quantity. At present, however, one may state that the determinations of precession from the lunar data show no significant difference from the value determined by Fricke.

Once the precession correction has been adopted, the motion of the equinox is determined from the secular part of the right ascension residuals. This value may also be somewhat subject to observational errors of the daytime objects, but it is not probable that such errors could dominate the solutions. The equinox motion was determined in the solution of DE111. For DE118, on the other hand, Fricke's value for E was also adopted and a plot of the right ascension residuals versus time indeed shows a general slope of about -0".5/cty, in agreement with the difference between the DE111 value for E and that of Fricke's.

At present, then, it seems that the ephemeris solutions differ from Fricke's determination of the equinox offset by about 0".5/cty. However, this difference may well be within realistic error bounds. Perhaps an analysis of the motion of Mars alone would be warranted, as this is the only non-daytime planet whose motion is determined by ranging.

B. Equinox Location

The 0-point of the ephemerides is theoretically that of the FK4; locating the dynamical equinox then gives the equinox offset. Good agreement is found for these values which indicate by how much the 0-point of the FK4 fails to coincide with the dynamical equinox. The range of the differences in the values are consistent with the expected uncertainties in both methods, about ± 0".05.

C. Obliquity

The instantaneous obliquity is well-determined by lunar laser-ranging which is highly sensitive to both the orientation of the spinning earth and to the plane of the perturbing solar orbit. One may then compute a value for this quantity directly from the ephemerides themselves (see, e.g., Standish 1982b). The ephemeris values are highly consistent; they are expected to be accurate to 0".01 or less. The standard (IAU) value is seen to differ from them by 0".02.

Table I. Comparisons of quantities determined by the planetary and lunar solutions with those determined by Fricke. The differences between the DE111 and DE118 values tend to indicate the uncertainties in the ephemeris determinations.

QUANTITY	FRICKE/IAU	DE111	DE118
Precession Correction	$1''.1$/cty	[$\doteq$ $1''.1$/cty]	[$\doteq$ $1''.1$/cty]
Equinox Motion	$1''.275$/cty	$1''.792$/cty	[$\doteq$ $1''.275$/cty]
Equinox Location	$0''.525$	$0''.587$	$0''.531$
Obliquity (- 23°26'44")	$0''.836$	$0''.816$	$0''.816$

III. CONCLUSIONS

Methods other than the traditional ones involving optical observations are beginning to produce determinations of a number of astronomical quantities. Planetary and lunar ephemerides, lunar laser-ranging, satellite ranging and VLBI are all capable of determining values for the orientation of the earth. However, at present, there are no indications of any significant changes to the values determined by Fricke; a slight correction to the IAU value of obliquity is indicated.

REFERENCES

Chapront-Touze,M.,Chapront,J.:1983, Astron.Astrophys. 124, 50.
Fricke,W.:1971, Astron. Astrophys., 13, 298.
1982, Astron. Astrophys., 107, 113.
Newhall,XX,Standish,E.M.,Williams,J.G.:1983, Astron.Astrophys.,125, 150.
Standish,E.M.:1973, Astron.Astrophys., 26, 463.
1982a, Cel.Mech., 26, 181.
1982b, Astron.Astrophys., 114, 297.
1985, Proceedings of IAU Colloquium #109, H.K.Eichhorn ed., Reidel, Dordrecht.

Fundamental Coordinate Ties Using Laser Ranging Data

Peter J. Shelus
McDonald Observatory and Department of Astronomy
University of Texas at Austin
Austin, Texas 78712
USA

ABSTRACT. One of the main goals of fundamental astronomy is to allow us to locate ourselves accurately in space and time. Each observational technique cannot supply all of the answers all of the time, and the joining of results from different techniques requires a true understanding of the particular reference frame which underlies each one. In this paper we try to state the present situation in our attempts to tie artificial satellite and lunar laser ranging data through the construction of stations designed to make both kinds of observations. This type of development strengthens both data types and eases their combination with other data types.

1. Introduction

One of the major goals of fundamental positional astronomy is to allow us to locate ourselves accurately in space and in time. Over the past several decades, almost unbelievable strides forward have been taken in both the theoretical and the observational arenas in an attempt to realize this specific goal. A number of these fundamental strides are being acknowledged by several excellent papers which are being presented here during this very meeting.

At the present time, more than ever before, the very small scale motions of our observing platform, the Earth, must be taken into consideration before its large-scale motions can be completely understood. It is only then that we can begin our search for a better and better realization of a satisfactory approximation to an inertial system of reference. A few of these new considerations involve plate tectonic motion (on both the intra-plate as well the inter-plate scales), the three-dimensional rotation of the Earth, and solid body tides, as well as the more familiar effects of refraction, aberration, precession, and nutation. To paraphrase from Ivan I. Mueller's fine paper *Reference Coordinate Systems for Earth Dynamics* presented at **IAU Colloquium No. 56** in Warsaw in September 1980, we can say that two basic reference systems are needed. The first is the inertial system, which, operationally, is probably to be attached to the extragalactic celestial radio sources. This system serves as a reference for the motion of the second system, a terrestrial one, which moves in some average sense with the earth and is attached to it through a number of dedicated observatories which operate on the earth's surface.

Celestial Mechanics **37** (1985) 243–246.

2. The Case for Laser Ranging

The concept of laser ranging to a celestial target entails the firing of a very narrow pulse of coherent photons from a station on the Earth to a reflector in Earth orbit, noting the returning signal, and recording the total time of flight. Those observations can reveal as much information about the dynamics of the site from which the observations are being made as about the dynamics of the target itself. Such observations, using retro-reflectors on many different artificial satellites and the Moon, have been a more or less common event since the mid to late 1960's. Both artificial satellite and lunar laser ranging has been continuously contributing to our knowledge of the motions of the Earth. Most recently these contributions have come about as a part of the efforts connected with NASA's Crustal Dynamics Project and the international MERIT (Monitoring of Earth Rotation and the Inter-comparison of Techniques) campaigns.

With respect to coordinate frames, the main references for the lunar system are the orbital planes of the Moon and the equator of the Earth. The accuracy with which this lunar system can be realized is determined by the accuracy of our knowledge of the lunar orbit and on how well the lunar librations can be modeled. The lunar ephemeris (both orbital and rotational) is presently most accurately determined by a consideration of lunar laser ranging observations and is generally obtained through the process of numerical integration. The dependence of these results upon modeling and on the choice of initial conditions is obvious. And, ultimately, the question reduces to how well this particular coordinate frame can be compared to other frames. A connection to the stellar frame might be made through the occultations of stars by the Moon; a connection to the radio source frame might be made through differential VLBI observations between radio source packages deposited on the surface of the Moon and standard extragalactic radio sources; a connection to the planetary frame might be made through the planetary ephemeris which is used in calculating the lunar ephemeris.

In the case of a reference system defined by ranging observations to artificial satellites, we find that the conditions are, as might be expected, almost identical to those which we meet when dealing with the lunar situation. There are certainly differences, however. On the positive side, the physical processes of actually laser ranging to an artificial satellite are several orders of magnitude more straightforward than those of laser ranging to the Moon. Further, because of the faster "mean daily motion" of an artificial satellite when compared to the Moon, one does not have to observe over as long a period of time in order to secure satisfactory differences in observing conditions for the maximum separation of variables. On the negative side, there are additional modeling problems related to the near-Earth gravitational field and its effects on the fine-scale motion of the artificial satellite. And, there are no direct ties to other frames of reference. The main reference planes, the orbital plane of the satellite and the Earth's equator, intersect along a "line of nodes", the initial orientation of which must me defined more or less arbitrarily. The accumulation of errors in describing the motion of the node with respect to the selected zero point is another problem.

3. Dual-Purpose Laser Ranging Stations

Within the NASA Crustal Dynamics Project and the interested international community, there has been evolving the concept of dual-purpose artificial satellite and lunar laser ranging stations. Two such stations have just recently begun such "simultaneous" observations. One is located at the LURE Observatory atop Mount Haleakala on the island of Maui in Hawaii, operated under NASA contract by the Institute for Astronomy of the

University of Hawaii. The other is the McDonald Laser Ranging Station, MLRS, located at McDonald Observatory near Fort Davis, Texas, and operated, also under NASA contract, by the Department of Astronomy of the University of Texas at Austin. A third station is in the final stages of development by the Division of National Mapping in Australia, and is located at a site in the Orroral Valley just south of Canberra. This station has already begun to operate successfully in the artificial satellite mode and is scheduled to become lunar capable within the next few months. A fourth such station is presently in the planning and initial construction stages and is scheduled to become operational at the Wettzell site in the southeastern region of the Federal Republic of Germany within the next two years. A fifth station, at the present, lunar capable only, is in routine operation at CERGA, near Nice in France. For the past two years or so, significant amounts of good lunar data have been produced there, and a major upgrade of the system is planned in the very near future.

Each of these stations are equipped with sub-nanosecond pulse lasers in both the lunar and artificial satellite mode. Because all of the equipment for both modes of observing are contained within the system, it is a very straightforward procedure to go from artificial satellite mode to lunar and vice-versa, often in time scales of minutes (if not seconds). Near each of these laser ranging facilities there is observing equipment for at least one other technique, i.e., optical, Doppler, VLBI, etc. Thus, the opportunity exists to obtain data of the first quality in both artificial satellite and lunar mode and then to compare and combine them with each other as well as with data from the other techniques.

4. Data Production and Reduction

For a geophysically meaningful reduction, it is convenient to define the topocentric distance, ρ, between an observing station and a retroreflector target in Earth orbit as approximately

$$\rho = R_1 - \sigma \cos \delta \cos H - z \sin \delta$$

where R_1 is the geocentric distance to the target; σ is the spin axis distance of the station; z is the equatorial distance of the station; δ is the declination of the target; and H is the local hour angle of the target. Theoretically of course, there is no restriction forcing one to work in this cylindrical coodrdinate system for station position; one could just as well work with spherical, rectangular, or any other covenient reference system for station coordinates. Applying an observational equation of the form

$$\rho_o - \rho_c = \sum_{i=1}^{n} \frac{\partial \rho_c}{\partial \kappa_i} (\kappa_i - \kappa_c)$$

and using standard methods for this kind of dynamical parameter improvement, we can obtain an observational equation for deriving "improvements" to the station position to be as follows:

$$\Delta\rho = [-\sigma \cos \delta \sin H]\, \Delta\lambda + [-\cos \delta \cos H]\, \Delta\sigma + [-\sin \delta]\, \Delta z.$$

Reducing our observations in this way, one solves for station location within the reference frame of each of our models, which, in the case of these dual-purpose laser ranging stations, can then be directly compared to produce offset parameters tying the two reference systems to one another.

5. Conclusions and Comments

All of that which has been presented here is both straightforward and obvious. It depicts a procedure which has been used again and again to relate one system of observations to another. In this specific case we are attempting to tie two reference frames together through observations in those two frames which are made by exactly the same instrumentation. It should have been done long ago in the past, except for the problems of having a operating network of such dual-purpose stations available. Although we have been striving toward this goal for almost 15 years now, the end is in sight in that stations are really now in operation.

SATELLITE LASER RANGING AND ITS APPLICATIONS

B. D. Tapley, B. E. Schutz and R. J. Eanes
The University of Texas at Austin
Center for Space Research
Austin, Texas 78712
U.S.A.

ABSTRACT. Satellite laser ranging (SLR) provides an important capability for precise orbit determination and for geophysical parameter estimation to support a number of contemporary geodynamic and oceanographic investigations. The precision of the SLR measurement has improved from the early meter-level systems to the current capabilities of a few centimeters for the best systems. The accuracy of the orbits and geophysical parameter recovery have shown an associated improvement. Polar motion with accuracies of 2 mas, station coordinates better than 10 cm, and interstation baseline rates indicative of tectonic motion are determined routinely with the current set of global SLR data. This discussion reviews the SLR measurement, analysis approach, and some of the recent results derived from the current SLR data set.

1. INTRODUCTION

Soon after the development of the laser, the ability to range to a near-earth satellite was demonstrated [*Plotkin et al.*, 1969]. Since then, laser ranging to cube-corners carried by near-earth satellites and to targets on the moon has evolved into an established technique for precise ephemeris determination. Results obtained from the SLR data provide the basis for a number of contemporary studies in geodesy, earth dynamics and lunar science [*Schutz et al.*, 1981; *Smith et al.*, 1976, 1985; *Yoder et al.*, 1983; *Robertson et al.*, 1985; *Tapley et al.*, 1980, 1985]. In addition to its contribution to geodesy and geodynamics, satellite laser ranging has played a major role in satisfying the orbit determination requirements of recent satellite altimeter missions [*Tapley et al.*, 1980b, 1982a]. As a consequence of the interest in these missions, a global network of laser satellite tracking systems has been established [*Coates et al.*, 1985].

Since an artificial satellite orbiting near to the earth is perturbed by the gravitational attractions of the inhomogeneous mass distribution of the earth, the time-varying effects of solid body and ocean tides and surface force interactions due to atmospheric drag and solar radiation pressure, the computation of its orbit is quite complex. However, the sensitivity of the satellite's motion to these effects provides an opportunity to use the SLR data to improve our knowledge of them [*Eanes et al.*, 1983; *Felsentreger et al.*, 1979; *Lambeck et al.*, 1974; *Lerch et al.*, 1981]. Because of the greater distance, the lunar laser range (LLR) measurement is more difficult to perform and until recently, only one site has ranged regularly to the moon during the nearly two decades since the reflectors were placed on the lunar surface by the Apollo astronauts. Nevertheless, the results obtained by both SLR and LLR have led to significant advances in our knowledge of the motions of the earth and the moon and in the fundamental constants required for the description of these motions. See, for example, *Cappallo et al.*, 1982; *Dickey and Williams*, 1982; *Ferrari et al.*, 1980; *Goad and Douglas*, 1978; *Yoder et al.*, 1978; *Langley et al.*, 1981a; *Williams et al.*, 1976; *Yoder et al.*, 1983.

In the following discussion, attention will be devoted to satellite laser ranging to artificial satellites and to some of the recent geophysical results obtained by analysis of these measurements.

Celestial Mechanics **37** (1985) 247–261.

2. SATELLITE LASER RANGING

Satellite laser ranging has been used to track the cube corner reflectors carried on Beacon Explorer-C, Geos-3, Seasat, Starlette, Lageos and several other satellites. Two of these satellites, Geos-3 and Seasat, carried altimeters which required orbit accuracies commensurate with the precision of the altimeter measurement for the analysis of the data. For Seasat in 1978, this required an orbit whose radial orbit component was accurate to ten centimeters [*Tapley and Born*, 1980].

The altimetric satellites, along with the launch of Lageos in 1976, initiated a major international activity in the application of SLR to satellite geodesy. Under NASA's Crustal Dynamics Project [*Flinn*, 1981; *Coates et al.*, 1985], the laser systems have been steadily improved in accuracy to a few centimeters [*Tapley et al.*, 1982; *Tapley et al.*, 1985]. During the last few years, a cooperative network of NASA stations and non-U.S. cooperating stations has evolved with a global distribution of the sites. These data have been used for the determination of precise satellite orbits, the earth's gravity field, the earth's polar motion and rate of rotation, solid earth and ocean tide parameters, planetary rehology and tectonic plate motion and crustal deformation. The goals set forth in several current international global earth dynamics programs, including the NASA Crustal Dynamics Project [*NASA Office of Space and Terrestrial Applications*, 1979; *NRC Committee on Geodesy*, 1978], the San Andreas Fault Experiment and the IAU/IUGG MERIT Campaign [*Wilkins*, 1980], impose significant challenges for the laser network observation efficiency and accuracy.

As an example, the objectives of the Crustal Dynamics Project are to study the motions of the earth's tectonic plates and the overall dynamic motion of the earth with the objective of improving the knowledge of

1. Regional deformation and strain accumulation related to large earthquakes in the plate boundary regions, such as the western North America region.
2. The contemporary motions of the North American, Pacific, South American, Nazca, European and Australian Plates.
3. The internal deformation of continental and oceanic lithospheric plates, with particular emphasis on North America and the Pacific.
4. The rotational dynamics of the earth and their possible correlation to earthquake, plate motions and other geophysical phenomena.

Table 1 summarizes the geophysical measurement accuracy required to achieve the objectives of the Crustal Dynamics Project [*NASA*, 1979].

TABLE 1. GEOPHYSICAL MEASUREMENT ACCURACY FOR THE CRUSTAL DYNAMICS PROJECT

Crustal Motions				
	σ_v, cm/yr	ΔT, yr	σ_d, cm	σ_n, cm
Current	2	5	11	8
Near future	1	5	6	4
Goal	.4, .7	5, 3	2	1

σ_v, baseline rate of change determination accuracy
ΔT, measurement span
σ_d, position determination accuracy
σ_n, laser range normal point accuracy

Earth Rotation Parameters			
	x_p, y_p; mas	UT1, ms	LOD, ms/D
Current	5	5	3
Near future	2	2	1
Goal	1	1	.5

The following discussion reviews recent improvements in the Lageos laser ranging data with regard to both the improvement in data quality and the geographical distribution of the data. Finally, recent results obtained from satellite laser ranging are reviewed.

3. THE SATELLITES

The primary target for geodetic applications has been the Laser Geodynamics Satellite (Lageos) [*Johnson et al.*, 1976; *Fitzmaurice et al.*, 1977]. Lageos was launched in May 1976 into a 5900 km altitude, 110° inclination and near-circular orbit. It is a 411 kilogram, 60 centimeter diameter aluminum sphere with a beryllium-copper core, covered with 426 retroreflectors. Four of the retroreflectors are germanium, for use with infrared lasers, and the remainder are fused silica for use with optical wavelength lasers. The high altitude of the Lageos orbit combined with the satellite's spherical shape and large mass diminish the influence of errors in the gravitational and the nongravitational force models.

After Lageos, the next most important satellite for laser ranging has been the satellite Starlette which was launched by the French Centre National d'Etudes Spatiales (CNES) [*CNES*, 1975]. The orbit inclination is 49.8°, and the perigee and apogee altitudes are 810 km and 1105 km, respectively. As with Lageos, both the satellite construction and orbit were designed to minimize the effects of surface forces, e.g., drag and solar radiation pressure. Starlette has a radius of 12 cm and a mass of 47.295 kg. The low area-to-mass ratio was achieved by using depleted uranium for the core. The shell is an aluminum alloy on which 60 laser cube corner reflectors were placed.

Other satellites which have been tracked for geodetic purposes include Geos-3, Beacon-C and Seasat. While tracking of these satellites has produced useful results, they have not been as interdisciplinary in nature as Lageos and Starlette, and primary emphasis in this discussion will be results obtained by laser ranging to these two satellites.

4. THE LASER RANGE MEASUREMENT

A satellite laser ranging station consists of a laser and transmitting and receiving telescopes mounted on a suitable tracking mount and set up on a site marked with a permanent marker [*Fitzmaurice*, 1978]. The laser transmits a pulse of energy at a measured time toward a satellite equipped with optical cube corners on its earth-viewing side. The cube corner on the satellite reflects the incident laser pulse back to the laser receiver at the station. The return of the laser pulse in the receiving telescope is time-tagged, and the round-trip travel time of the pulse can then be determined. This process is controlled by the station computer which automatically starts and stops the laser firing and pulse counter. The pulse duration depends on the laser design and may vary from a few nanoseconds for the older systems down to a hundred picoseconds for the best systems.

For calibration purposes, the system ranges to a ground target at a known distance prior to and after the satellite tracking has been performed. During the calibration period, the satellite laser system ranges to the calibration target and the mean value of these measurements is adjusted to agree with the known distance. Errors in the survey measurement of the calibration target distance will lead to errors in the measured range. The possibility of physical shifts in the calibration target position with time as well as the effects of atmospheric refraction errors must be considered. To improve calibration accuracy, continuous calibration measurements over an internal path are now being made during satellite ranging with some of the newer systems.

Figure 1 illustrates the various elements in the laser range measurement.

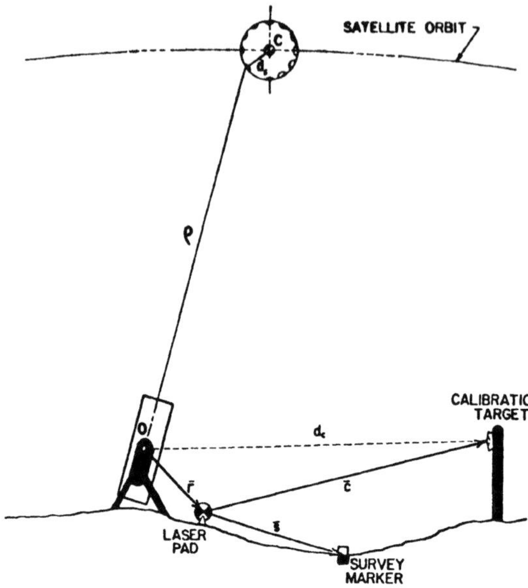

Figure 1. Satellite laser range measurement.

The primary measurement is the flight time, Δt, between the laser pulse transmission and the return of the satellite reflected energy to the optical receiver at the tracking station. With appropriate corrections for atmospheric refraction and instrument delays, the measured time of flight can be used to obtain the range from the first stationary point in the optical path of the laser (the reference point) to the average satellite corner-cube position at the time the pulse is reflected at the satellite. That is,

$$\rho = \frac{c\,\Delta t}{2} - \eta_r - \eta_e - b + \varepsilon \tag{1}$$

where

ρ is the range from the laser reference point to the average corner-cube reflector position.

Δt is the round-trip flight time.

c is the speed of light in a vacuum.

η_r is the atmospheric refraction correction.

η_e is the effect of systematic and random measurement errors.

b is the system delay as determined by calibration measurements.

ε is the unmodeled observation error.

Eq. (1) suggests that the errors which corrupt the measurement of the round-trip time of flight can be separated into instrument or measurement errors, errors in the models used to correct the time of flight for atmospheric refraction and other effects. Corrections for these effects require that ancillary data, such as atmospheric pressure, temperature and relative humidity be obtained [*Marini and Murray*, 1973].

The equations of motion predict the position of the satellite's center of mass; hence, the location of the effective optical center, d_s, (Fig. 1) of the laser corner-cube array with respect to the spacecraft's mass center must be known precisely [*Fitzmaurice et al.*, 1977]. Since the laser reference point, O, will vary as different instruments occupy a given site, the surveyed distance, $\bar{r}$, between the laser reference point and the laser site benchmark must be determined precisely for each new site occupancy. Finally, since some applications will require intercomparison of positions and baselines determined by various techniques, the location of the laser station benchmark with respect to a specified survey mark, $\bar{s}$, must be determined precisely.

The instrument errors, η_e, include errors in the time standard, which can be separated into clock bias (relative to UTC), drift and discontinuous or anomalous time standard behavior. It is generally assumed that the transmitted pulse is symmetrical in shape. A non-Gaussian pulse can lead to erroneous interpretation of time of flight for the returned signal unless corrections for this effect are included in the data preprocessing.

Another common error source occurs in the determination of the system delays from pre- and post-observation calibration measurements. Factors which influence the overall accuracy of the bias measurement are uncertainties in the measurement of the calibration target distance, d_c, background interference in ranging to the target, atmospheric instability in the calibration medium and differences in the signal strength of calibration versus satellite returns.

5. THE LAGEOS DATA SET

The tracking stations which participated in the MERIT Campaign [*Wilkins*, 1980] are shown as black circles in Figure 2. The performance of the various systems are summarized in Table 2, which gives the average full-rate and quick-look tracking precisions.

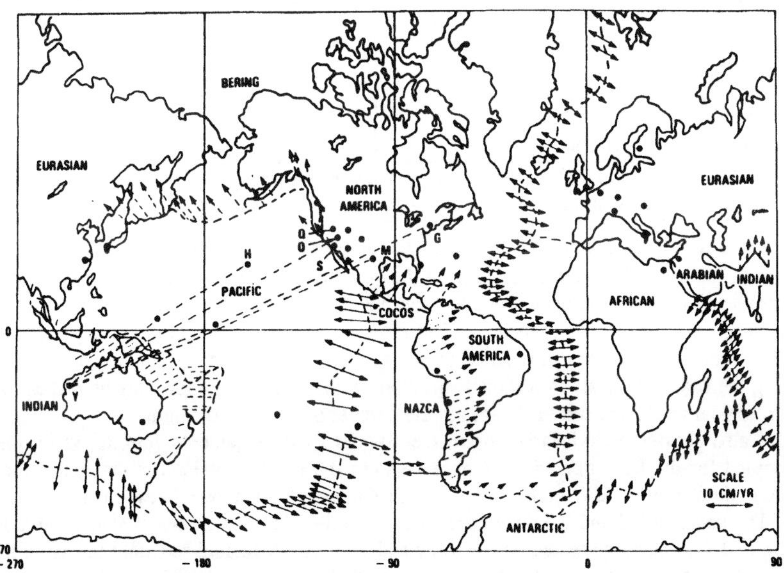

Figure 2. Tracking stations and motion at tectonic plate boundaries (adapted from *Minster et al.*, 1974).

Figure 3 shows the temporal distribution of the Lageos data set as well as the improvement with time in the tracking accuracy. There are obvious time periods when substantially more data is collected than at other

TABLE 2. SDS MONTHLY AVERAGE PRECISION (cm)

	FR	NP
1181 Potsdam	19	9
7086 Ft. Davis	8	1
7090 Yaragadee	2	1
7105 Greenbelt	3	1
7109 Quincy	3	1
7110 Monument Peak	3	1
7112 Platteville	12	3
7121 Huahine	10	2
7122 Mazatlan	10	1
7210 Haleakala	4	1
7265 Mojave	8	1
7400 Santiago	7	1
7401 Cerro Tololo	8	1
7805 Metsahovi	23	17
7810 Zimmerwald	8	2
7831 Helwan	21	10
7833 Kootwijk	15	7
7834 Wettzell	7	1
7835 Grasse	6	1
7837 Shanghai	17	12
7838 Simosato	9	1
7839 Graz	4	1
7840 Herstmonceux	5	2
7843 Orroral	6	4
7886 Quincy/TLRS-1	7	1
7907 Arequipa	14	4
7939 Matera	14	3
7940 Dionysos	16	11
8833 Kootwijk/MTLRS-1	5	2

Note: The minimum NP precision estimate was allowed to be 1 cm to avoid unrealistically small estimates.

periods, and some stations have contributed a much larger number of observations than other stations.

A number of factors influence the variation in the quantity of data collected at any given station. The dominant variation in the data collected at each site has a period of 560 days and is associated with the motion of the Lageos node relative to the sun. This effect is due to the ratio of daytime passes to nighttime passes at each tracking site and is introduced by the limited ability of many tracking stations to range to Lageos during the daylight hours. Additional variations with an annual period may be present due to seasonal meteorological variations and to changes in the number of hours of daylight.

There are also shorter period variations in the amount of data gathered introduced by the tracking schedule adopted for the laser operation. Figure 4 shows the results obtained by analyzing the data collected each day during the 92-day preliminary MERIT Campaign conducted between August 1, 1980, and October 31, The peaks at 7.0 and 3.5 days occur because a five-day work week is scheduled at most sites, leading to limited data being collected on Saturday and Sunday. A rotated schedule to ensure tracking by an adequate number of stations during the weekend could eliminate this schedule-induced variation in the tracking data. 1980.

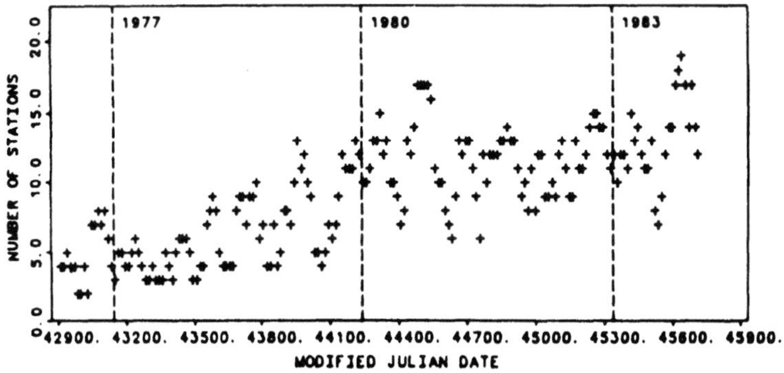

Figure 3*a*. Number of stations observing in each 15-day interval from May 1976 through January 1984.

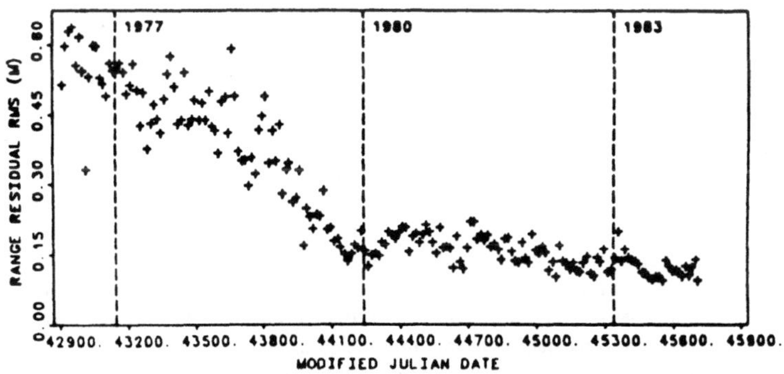

Figure 3*b*. Root-mean-square of laser range residuals (observed minus computed) after solution for station coordinates, earth orientation, and 15-day mean orbit element corrections. A point is shown for each of the 15-day orbital arcs.

6. METHOD OF ANALYSIS

The definition of an appropriate reference coordinate system is a basic requirement for the description of phenomena associated with the motion of any dynamic system. For solid earth dynamics studies, a body-fixed coordinate system is appropriate for the geophysical models, while an inertial frame is appropriate for describing the motion of the spacecraft.

The fundamental role of the terrestrial or body-fixed coordinate system will be to locate the tracking stations and to identify and study the evolution of any time-dependent phenomena related to changes in the size and shape of the earth. The satellite laser measurements can determine the location of a set of tracking stations in an appropriate terrestrial coordinate system. After establishing the tracking station locations, the motion of the earth's rotation axis relative to this system can be studied. However, separation of changes associated with the angular velocity vector and those associated with tectonic motions of the tracking stations is a potential problem.

In the approach used at The University of Texas Center for Space Research, 7.7 years of Lageos laser ranging data is used to adjust a single set of station positions simultaneously with orbit and earth rotation

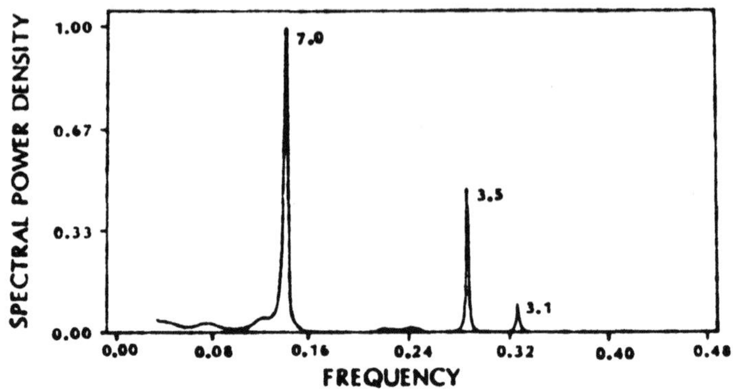

Figure 4. Maximum entropy spectral analysis of Lageos observation distribution.

parameters. This global station coordinate solution provides the basis for an initial definition of the terrestrial reference system. In a subsequent solution, tracking station positions are adjusted using successive 60-day intervals, and the time-ordered change in these coordinates is compared to the change predicted by a model of tectonic plate motion. Finally, the earth rotation parameters are determined as an inherent part of the solution process.

In the approach adopted in this investigation, the University of Texas orbit analysis system, UTOPIA, was used to obtain a dynamically consistent solution to 7.7 years of Lageos laser range data. The data set spanned May 7, 1976, to January 17, 1984, and included 130,439 normal points, constructed from three-minute averages of the laser range observations from 57 different laser ranging sites. The normal point precision of the various systems varied from about a meter for some systems in 1976 to better than 50 mm for many of the systems at the end of the solution interval. These data were used to obtain an accurate long arc orbit solution by simultaneously adjusting the satellite's initial position and velocity along with 40 coefficients in a Fourier series approximation to the unmodeled along-track acceleration, 10 constituents of the model for the dynamical effect of solid-earth and ocean tides [*Eanes et al.*, 1983], and the zonal gravity field coefficients J_2, J_3, J_4 and J_5. The values of C_{21} and S_{21} were fixed to the values implied by the mean BIH polar motion during the years 1972-1978 [*Melbourne et al.*, 1983]. The remaining gravity field coefficients were set to the GEM-L2 values [*Lerch et al.*, 1985]. For the 7.7-year data set, the best-fitting dynamically consistent solution, denoted as LLA 8402, resulted in a root-mean-square observation residual of 1.07 m. The residuals are dominated primarily by long-period systematic errors in the orbit elements due to errors in the force model.

Although the 1.07 m fit of the range observations during the 7.7-year period represents an accurate initial ephemeris, it will not support the solution of baselines and tracking stations with accuracies at the level of a few centimeters. To remove the effects of the remaining orbit and earth orientation error, the 7.7-year solution was separated into 15-day intervals. In each 15-day interval, the LLA 8402 observation residuals were used to solve for a constant correction to the orbit elements implied by the long-arc solution. The node was held fixed to the long arc solution and the remaining elements were adjusted with no a priori information assumed. Within each 15-day interval, corrections to the nominal values of polar motion and universal time (UT1) were determined at three-day intervals. A single solution for each tracking station was determined for the complete 7.7-year data span. To constrain the longitude origin, which is not directly determined by the laser range measurements, the x and y coordinates of the site at McDonald Observatory were adjusted with an a priori standard deviation of 50 mm. The range residuals, after applying the orbit, earth orientation and station coordinate corrections, are substantially improved over the long-arc residuals.

7. GEOPHYSICAL AND GEODETIC RESULTS FROM SATELLITE LASER RANGING TRACKING STATION SOLUTIONS

The Lageos laser range data collected over the 7.7-year period since the launch of Lageos has been used to determine the coordinates of forty independent tracking locations. In the solutions, the tracking station coordinates are held constant over the data interval. Consequently, the average value of the station coordinates for a seven-year interval will be determined with the effects of tectonic plate motion averaged into the solution. Solutions for the station coordinates, over shorter averaging intervals, can be used to determine the relative separations referred to as baselines. Successive coordinate solution determinations can be used to determine time rates of change in the baselines as an indicator of tectonic motion.

The accuracy of recent solutions for the global tracking station coordinates is better than 10 cm [*Smith et al.*, 1983; *Tapley et al.*, 1985]. Table 3 compares the coordinates of 26 tracking stations which were independently determined in the SL5 solution by *Smith et al.* and the LSC8402 solution given by *Tapley et al.* [1985].

TABLE 3. LAGEOS STATION COORDINATES (LSC8402) COMPARED TO THE GODDARD SPACE FLIGHT CENTER SL5 SOLUTION

Station		*Differences Before Fit*			*Differences After Fit*		
		x (m)	y (m)	z (m)	x (m)	y (m)	z (m)
QUN081	7051	.075	−.105	−.440	.018	−.001	−.074
TLROTA	7062	.009	−.089	−.436	.010	−.056	−.031
STALAS	7063	−.139	−.253	−.576	−.066	.014	−.019
BDALAS	7067	−.025	−.280	−.540	.103	−.007	.050
RAMLAS	7069	.081	−.234	−.560	.223	−.085	.027
TLRBLK	7082	.125	−.243	−.472	.094	−.090	−.052
MCDXAA	7086	−.032	−.050	−.487	.032	.010	.002
YARLAS	7090	.264	.264	.594	.005	−.024	−.046
HAYST2	7091	−.131	−.348	−.568	−.079	−.025	−.022
KWALAS	7092	.250	.157	.329	−.073	−.053	−.017
SAMOA1	7096	−.095	.620	.325	−.197	.119	.123
QUN81B	7109	.140	−.123	−.403	.083	−.019	−.037
MNPEAK	7110	.012	.031	−.438	.013	.069	−.029
PLATVL	7112	−.053	−.059	−.451	−.051	.098	.010
OWNSV2	7114	.021	−.076	−.400	−.010	.005	−.012
GLDST2	7115	.013	−.081	−.428	−.000	−.018	−.026
HAWXAA	7120	.132	.125	−.045	−.025	−.040	.002
KOOTWK	7833	.194	−.748	−.176	.032	−.111	.046
GRASSE	7835	.114	−.781	−.299	−.015	−.161	−.081
TLRFLG	7891	.017	.036	−.420	.025	.116	.016
TLRVER	7892	−.012	.012	−.510	−.027	.154	−.075
TLRJPL	7896	−.063	−.036	−.416	−.077	.011	−.021
ARELAS	7907	−.430	.196	−.486	−.067	−.037	.038
HOPLAS	7921	.003	−.018	−.376	.034	.026	.070
NATLAS	7929	−.208	−.010	−.251	.085	.082	.234
ORRLAS	7943	.095	.535	.445	−.072	.025	−.077
		RMS Difference Before			*RMS Difference After*		
		.144	.305	.436	.079	.073	.067

The stations are compared by applying a seven-parameter least squares fit of one set of coordinates to the other set. In the fit, rigid-body translation parameters, DX, DY, DZ, a scale parameter and three rigid-body rotations were determined. These parameters imply a rigid-body translation of less than 6 cm for each axis. The scale parameter of $-2.6 * 10^{-8}$ is thought to be due to the different value of GM_e used for the two solutions. The value used for LSC8402 was $GM_e = 398600.448$ km^{3}sec^2, while the value used for the SL5 solution was $GM_e = 398600.434$ km^{3}sec^2. The LSC 8402 value is the adopted value for the

MERIT Campaign [*Melbourne et al.*, 1983], a value based on lunar laser ranging. As shown in the second part of Table 3, the internal precision implied by the rms agreement after the seven-parameter solution has been applied is less than eight centimeters.

Further insight into the scale of the satellite laser ranging solution can be obtained by comparison with independent length determinations from a comparably precise measurement process. There is a set of 22 baselines between stations on the North American Continent which have been determined by both laser tracking and VLBI measurements. Figure 5 shows the difference between the baseline lengths implied by the LSC8402 solution and the baseline lengths measured by the VLBI technique.

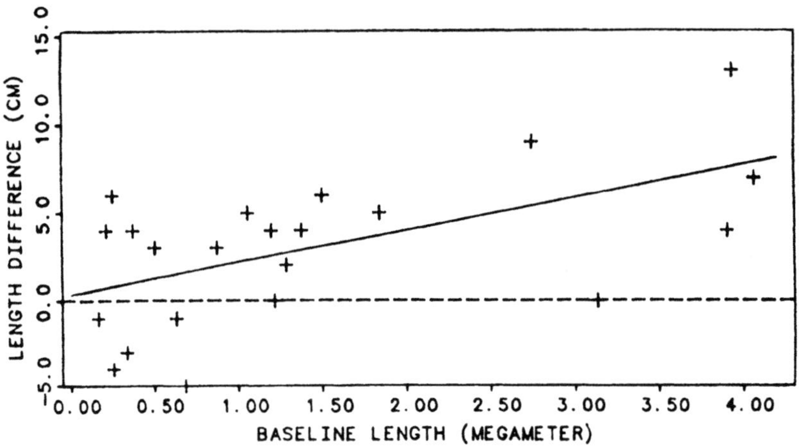

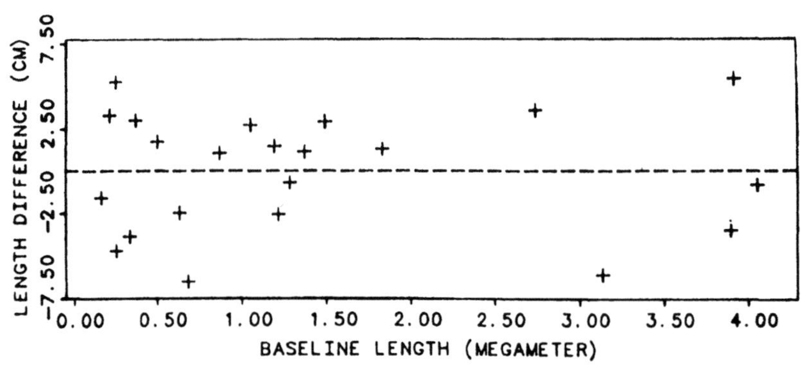

Figure 5. Differences (Lageos minus VLBI) in baseline length for 22 Crustal Dynamics Project intercomparison baselines, (a) before and (b) after removing a scale difference of 1.9×10^{-8}.

The scale of the SLR solution depends on the adopted values of both the speed of light and the value of GM_e. The VLBI solution will be dependent on the speed of light only. The best-fitting line to the differences shown in Figure 5 will have a slope of $1.9 * 10^{-8}$. When the SLR solutions are corrected for this apparent scale discrepancy, the rms agreement in the 22 baselines is 3.5 cm. In a subsequent solution, the value of GM_e was adjusted to obtain the value 398600.440. This value removes the scale difference of 1.9×10^{-8} between the SLR and VLBI solutions.

8. BASELINES AND BASELINE RATES

To determine a measure of tectonic motion [*Minster and Jordan*, 1978], individual tracking station coordinate solutions were determined in 60-day intervals. From the stations which tracked in each 60-day interval, baseline length (hereafter referred to as simply baseline) can be determined as the magnitude of the vector difference between these two stations. The baseline changes of most interest, tectonically, are those due to motion in the horizontal directions. As a consequence, in attempting to relate the successive baseline determinations to the motion predicted by current plate tectonic models, the height or vertical component of the individual tracking station solutions was constrained to the long arc solution value and only the latitude and longitude displacements were considered. By repeated determinations of the baselines, using tracking station solutions from successive 60-day intervals, changes in the baseline can be inferred. *Tapley et al.* [1985] show baseline changes between Yaragadee, Australia, which tracked continuously throughout the four year period considered here, and several stations on the North American plate and Pacific plate.

Figure 6 shows the baselines obtained by using the Yaragadee-Hawaii tracking station combination.

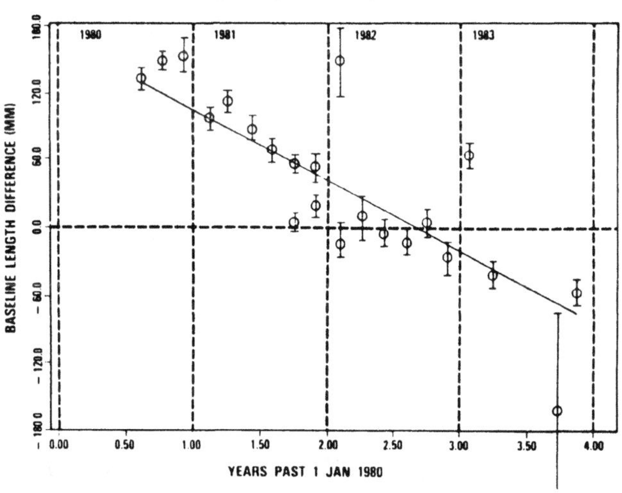

Figure 6. 60-day Lageos arcs. Yaragadee–Hawaii (Hollas and Moblas-1). Slope = −63 ± 3 mm/yr, (M&J = −67), rms = 32 mm.

The values shown are the baseline lengths, relative to the mean value, computed from the tracking station coordinates in successive 60-day intervals, during the period 1980 through 1984. The best fitting straight line has a slope of −63 (±3) mm/year. The value predicted by the Minster-Jordan AM1-2 model [*Minster and Jordan*, 1978] is −67 mm/year.

9. OTHER GEOPHYSICAL QUANTITIES

Other results of geophysical interest can be obtained from the Lageos data. During early analysis of the Lageos orbit, the observed along-track acceleration was found to be one or two orders of magnitude larger than expected, leading to a 1 mm/day decay in the semimajor axis [*Smith et al.*, 1985]. Considerable attention has been given to this effect in an effort to explain its origin. Early calculations suggested that charged particle drag was the principal cause. Recently, asymmetry in the earth's albedo radiation has been proposed as a candidate for contributing to this orbital behavior. A continued monitoring of the evolution of Lageos will be necessary along with further study of the upper atmospheric physics, and the models for the earth's albedo will be required to fully explain this effect.

Geodetic satellites whose orbits have been determined by laser tracking include Lageos, Starlette, Geos-3, Beacon-C and Seasat. In addition, laser tracking is proposed for several future missions including Lageos-II, Topex, Poseidon and ERS-1. Laser data from numerous satellites including Lageos have been incorporated in the recent models for the earth's gravity field [*Lambeck and Coleman*, 1983]. GEM-L2, which is the latest "satellite-only" gravity field derived by the NASA Goddard Space Flight Center, has a long wavelength geoid (to degree and order 4) which is assessed to be accurate to ±8 cm [*Lerch et al.*, 1982].

The SLR system has provided accurate measurements for orienting the earth within a quasi-inertial reference system. Global data taken over 5-day intervals provide a suitable means for solving directly the earth's polar motion and excess length of day [*Tapley*, 1983]. SLR solutions (for data taken since May 1976) are routinely provided to BIH at 5-day intervals for use in the *BIH Rapid Polar Motion Service* series. Annual solutions based on the full-rate data which include solutions for UT1 as well are reported in the *BIH Annual Reports* [*Schutz et al.*, 1984; *Smith et al.*, 1983]. Current accuracies, which are routinely achieved for 5-day mean values, are 2 milliarcsec for the X and Y coordinates of the axis of the rotation axis and 0.1 msec for change in length of day [*Robertson et al.*, 1985; *Schutz et al.*, 1985; *Tapley et al.*, 1985]. Figure 7 shows the earth rotation solutions for SLR during the period from 1980 through 1984. The black diamonds are independent determinations from the VLBI systems which did not begin regular reporting until the beginning of the MERIT Campaign in September 1983.

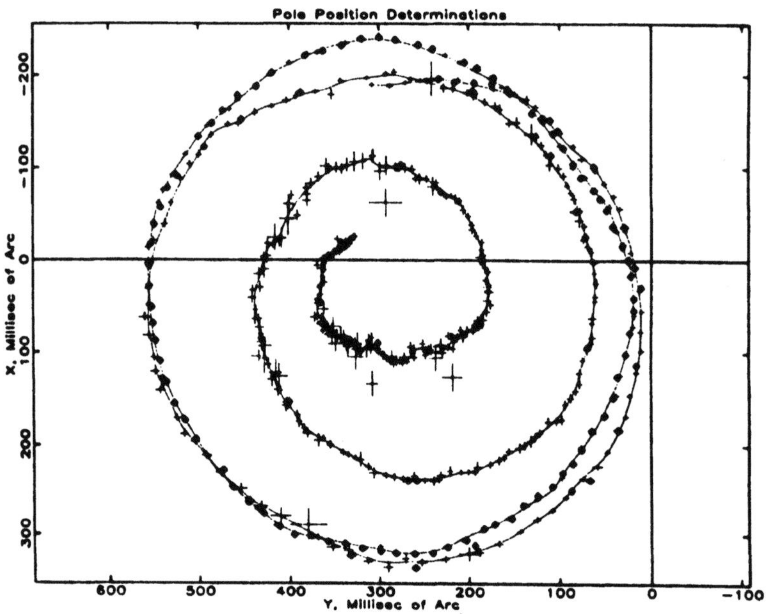

Figure 7. SLR and VLBI polar motion solutions for period from September 1980 through December 1983 (+ SLR; ♦ VLBI).

The analysis of the long period changes in the node have been interpreted in terms of the secular and seasonal changes in the earth's gravitational field coefficient J_2, the term which describes the earth's oblateness. The secular change in his value has been related to the viscoelastic rebound of the North America, Northwestern Europe and the Antarctica following the melting of large continental ice sheets subsequent to the last glacial maximum [*Yoder et al.*, 1983; *Peltier and Wu*, 1983; *Rubincam*, 1984; *Wu and Peltier*,

1984; *Lambeck*, 1980].

Comparisons of the Lageos-derived length of day (LOD) with the LOD derived from the exchange of the angular momentum of the atmosphere with the earth [*Langley et al.*, 1981b; *Smith et al.*, 1985] has indicated a very high correlation between the two time series. Currently, an intense effort is underway to understand the origin of the differences between the two data sets and to further understand the mechanism whereby the momentum interchange between the atmosphere and solid earth occurs.

The other retroreflector-equipped satellites mentioned previously are in orbits which are closer to the earth. Consequently, they experience a much richer spectrum of gravity and tidal effects than those experienced by the high-altitude Lageos. Of special interest is the French-launched Starlette satellite which, like Lageos, is a dense, passive ball covered by retroreflectors. It was launched exclusively as a laser target for geodetic applications. Considerable laser data has been acquired on Starlette and are being analyzed to improve the models for the earth's gravity field and ocean tides [*Marsh et al.*, 1978, 1985; *Cheng et al.*, 1985].

10. FUTURE PROSPECTS FOR SATELLITE LASER RANGING

During the next decade, there will be a substantial improvement in both the accuracy and the distribution of the laser tracking network. New sites in Austria, Japan, China, Italy, Israel and England and sites in the USSR will provide improved geographical distribution of the SLR network. Densification of the permanent network will be made at sites visited on a yearly basis by transportable laser ranging systems. More frequent reoccupations of selected sites by the transportable systems will be made for the purpose of investigating the rigidity of local networks. Improvements in the laser system accuracies and in the model for the error sources will results in normal point precision at the 5 mm level, and 2 to 3 cm orbit determination accuracy can be anticipated. The development of daytime ranging capability for all systems will enhance the temporal coverage of the satellite's motion and allow more rapid and accurate recovery of earth rotation and tectonic motions.

The international commitment to continued laser ranging will be strengthened with the launch of new satellites. A new Lageos satellite, Lageos-II, is planned for launch in 1988 as a combined mission between NASA and the National Space Plan of the National Research Council of Italy. POPSAT, which has been proposed as an ESA mission, would have both microwave and laser tracking. EGP, a combined photographic target and laser retroreflector satellite, will be launched by the Japanese. Finally, Topex, a joint U.S.-French oceanographic mission to map the general ocean circulation, is proposed for launch in 1991 and will carry a satellite altimeter, the improved TRANET system, and a laser tracking system.

With three-dimensional positioning accuracy expected to be on the order of 1 cm, the permanent laser network will serve as a fiducial set of points distributed around the globe. This network of points has the potential for maintaining a global datum accurate at the 1 cm level. Continuous monitoring of this network will play a significant role in determining earth rotation parameters and in modeling tectonic motions, plate deformations and vertical station movements on a global scale. The study of intraplate deformation and stability will be enhanced through the three-dimensional determinations of SLR fiducial points.

With the global SLR datum, better than 1 mas polar motion, determined daily, is anticipated. Earth orientation parameters of this accuracy will have a significant impact on studies of the polar motion frequency structure, and more precise determinations of the amplitude of the response at the Chandler, annual, semiannual, and diurnal periods as well as the nature of the Chandler excitation can be contemplated. Finally, the effects of atmospheric and oceanic phenomena on the orientation of the earth and the core/mantle coupling will be amenable to study also. With the development and deployment of improved laser ranging systems and the launching of additional satellites to complement the improved ground-based tracking network, the next decade holds substantial promise for those disciplines that depend on satellite laser ranging.

11. ACKNOWLEDGMENT

This research was supported by the National Aeronautics and Space Administration under the Crustal Dynamics Project, Contract No. NAS5-28192.

12. REFERENCES

Cappallo, R. J., R. W. King, C. C. Counselman, III, and I. I. Shapiro, 'Evidence for lunar librations near resonance,' *Celest. Mech.*, **26**, 145, 1982.

Centre National d'Etudes Spatiales, Groupe de Recherches de Geodesie Spatiale, *Starlette*, Toulouse, France, Feb. 1975.

Cheng, M. K., R. J. Eanes, C. K. Shum, B. E. Schutz, and B. D. Tapley, 'Earth rotation solution from Starlette orbit analysis during the MERIT Campaign,' *Proc. of International Conference on Earth Rotation and the Terrestrial Reference Frame*, Columbus, Ohio, 1985.

Coates, R. J., H. Frey, J. Bosworth, and G. C. Mead, 'Space age geodesy: The NASA crustal dynamics project,' *IEEE Trans. Geosci. Remote Sens.*, **GE-23**(4), 358-368, 1985.

Dickey, J. O., and J. G. Williams, 'Geodynamical application of lunar laser ranging,' *Eos Trans. AGU*, **63**, 301, 1982.

Eanes, R. J., B. E. Schutz, and B. D. Tapley, 'Earth and ocean tide effects on Lageos and Starlette,' in *Proceedings of the Ninth International Symposium on Earth Tides*, edited by J. T. Kuo, E. Schweizerbart'she Verlagsbuchhandlung, Stuttgart, 1983.

Felsentreger, T. L., J. G. Marsh, and R. G. Williamson, 'M_2 ocean tide parameters and the deceleration of the Moon's mean longitude from satellite orbit data,' *J. Geophys. Res.*, **84**(B9), 4675-4679, 1979.

Ferrari, A. J., W. S. Sinclair, W. L. Sjogren, J. G. Williams, and C. F. Yoder, 'Geophysical parameters of the Earth-Moon system,' *J. Geophys. Res.*, **85**, 3939-3951, 1980.

Fitzmaurice, M. W., P. O. Minott, J. B. Abeshire, and H. E. Rowe, 'Prelaunch testing of the laser geodynamic satellite,' NASA **TP-1062**, Oct. 1977.

Flinn, E. A., 'Applications of space technology to geodynamics,' *Science*, **213**, 89-96, 1981.

Goad, C. C., and B. C. Douglas, 'Lunar tidal acceleration obtained from satellite-derived ocean tide parameters,' *J. Geophys. Res.*, **83**, 306-310, 1978.

Johnson, C. W., C. A. Lundquist, and J. L. Zurasky, 'The Lageos Satellite,' paper presented at the International Astronautical Federation XXVIIth Congress, Anaheim, Calif., Oct. 10-16, 1976.

Lambeck, K., A. Cazenave, and G. Balmino, 'Solid earth and ocean tides estimated from satellite orbit analyses,' *Rev. Geophys. Space Phys.*, **12**, 421-434, 1974.

Lambeck, K., *The Earth's Variable Rotation: Geophysical Causes and Consequences*, Cambridge Univ. Press, London, 1980.

Lambeck, K., and R. Coleman, 'The Earth's shape and gravity field: A report of progress from 1958 to 1982,' *Geophys. J. R. Astr. Soc.*, **74**, 25-54, 1983.

Langley, R. B., R. W. King, and I. I. Shapiro, 'Earth rotation from lunar laser ranging,' *J. Geophys. Res.*, **86**, 913, 1981a.

Langley, R. B., R. W. King, I. I. Shapiro, R. D. Rosen, and D. A. Salstein, 'Atmospheric angular momentum and the length of day: A common fluctuation with a period near 50 days,' *Nature*, **294**, 730, 1981b.

Lerch, F. J., B. H. Putney, C. A. Wagner, and S. M. Klosko, 'Goddard earth models for oceanographic applications (GEM 10B and 10C),' *Marine Geod.*, **5**(2), 145-187, 1981.

Lerch, F. J., S. M. Klosko, and G. B. Patel, 'A gravity model for crustal dynamics (GEM-L2),' *J. Geophys. Res.*, **90**(B11), 9301-9311, September 30, 1985.

Marini, J. W., and C. W. Murray, 'Correction of laser range tracking data for atmospheric refraction at elevations above 10 degrees,' NASA TM X-70555, 1973.

Marsh, J. G., F. J. Lerch, and R. G. Williamson, 'Precision geodesy and geodynamics using Starlette laser ranging,' *J. Geophys. Res.*, **90**(B11), 9335-9345, September 30, 1985.

Marsh, J. G., and R. G. Williamson, 'Precision orbit computations with Starlette,' *Bull. Geod.*, **52**(1), 7182, 1978.

Melbourne, W., R. Anderle, M. Feissel, R. King, D. McCarthy, D. Smith, B. Tapley, and R. Vicente, *Project MERIT Standards*, U. S. Naval Observatory Circular No. 167, Washington, D. C., Dec. 27, 1983.

Minster, J. B., and T. H. Jordan, 'Present-day plate motions,' *J. Geophys. Res.*, **83**(B11), 1978.

NASA Office of Space and Terrestrial Applications, *Application of Technology to Crustal Dynamics and Earth Research*, NASA Tech. Paper 1464, Aug. 1979.

National Research Council, Committee on Geodesy, *Geodesy: Trends and Prospects*, Assembly of Mathematical and Physical Sciences, 1978.

Robertson, D. S., W. E. Carter, B. D. Tapley, B. E. Schutz, and R. J. Eanes, 'Polar motion measurements: Subdecimeter accuracy verified by intercomparison,' *Science*, **229**, 1259-1261, September 20, 1985.

Schutz, B. E., B. D. Tapley, and R. J. Eanes, 'Earth rotation from Lageos laser ranging,' *BIH Annual Report for 1980*, D27-D29, Paris, 1981.

Schutz, B. E., R. J. Eanes, and B. D. Tapley, 'The SLR operational center series ERP(CSR)84L02,' *Proc. of International Conference on Earth Rotation and the Terrestrial Reference Frame*, Columbus, Ohio, 1985.

Silverberg, E. C., T. Cahill, and J. Dorman, 'Relative lateration across the Los Angeles Basin using a satellite laser ranging system,' *Bull. Geod.*, **56**, 329-340, 1982.

Smith, D. E., F. J. Lerch, J. G. Marsh, C. A. Wagner, R. Kolenkiewicz, and W. D. Kahn, 'Contributions to the national geodetic satellite program by Goddard Space Flight Center,' *J. Geophys. Res.*, **81**(5), 1006-1026, Feb. 1976.

Smith, D. E., D. C. Christodoulidis, P. J. Dunn, and M. H. Torrence, 'Geodetic parameters estimated from Lageos laser data, SL5,' Fifth Annual NASA Geodynamics Program Conference and Crustal Dynamics Project Review, Washington, D. C., 1983.

Smith, D. E., D. C. Christodoulidis, R. Kolenkiewicz, P. J. Dunn, S. M. Klosko, M. H. Torrence, S. Fricke and S. Blackwell, 'A global geodetic reference frame from Lageos ranging (SL5.1AP),' *J. Geophys. Res.*, **90**(B11), 9221-9234, September 30, 1985.

Stolz, A., and D. R. Larden, 'Seasonal displacement and deformation of the earth by the atmosphere,' *J. Geophys. Res.*, **84**(B11), 6185-6194, 1979.

Tapley, B. D., B. E. Schutz, and R. J. Eanes, 'Global station coordinate solution from Lageos laser ranging,' *Eos Trans. AGU*, **61**, 1980a.

Tapley, B. D., and G. H. Born, 'The Seasat precision orbit determination experiment,' *J. Astronaut. Sci.*, **XXVIII**(4), 315-326, 1980b.

Tapley, B. D., G. H. Born, and M. E. Parke, 'The Seasat altimeter data and its accuracy assessment,' *J. Geophys. Res.*, **87**(C5), 3179-3188, April 30, 1982a.

Tapley, B. D., B. E. Schutz, and R. J. Eanes, 'A critical analysis of satellite laser ranging data,' in *Laser Ranging Instrumentation*, vol. II, *Proceedings of the 4th International Workshop*, pp. 523-567, Geodetic Institute, Bonn University, Bonn, 1982b.

Tapley, B. D., Polar motion and earth rotation, *Rev. Geophys.*, **21**(3), 569-573, 1983.

Tapley, B. D., R. J. Eanes, and B. E. Schutz, 'UT/CSR analysis of earth rotation from Lageos SLR data,' *Proc. of International Conference on Earth Rotation and the Terrestrial Reference Frame*, Columbus, Ohio, 1985.

Tapley, B. D., B. E. Schutz, and R. J. Eanes, 'Station coordinates, baselines, and earth rotation from Lageos laser ranging: 1976-1984,' *J. Geophys. Res.*, **90**(B11), 9235-9248, September 30, 1985.

Wahr, J. M., 'Body tides on an elliptical, rotating, elastic, and oceanless earth,' *Geophys. J. R. Astronaut. Soc.*, **64**(3), 1981.

Wilkins, G. A. (Ed.), *Project MERIT: A Review of the Techniques to Be Used During Project MERIT to Monitor the Rotation of the Earth*, p. 77, Royal Greenwich Observatory, Herstmonceux, 1980.

Yoder, C. F., J. G. Williamson, J. O. Dickey, B. D. Tapley, R. J. Eanes, and B. E. Schutz, 'Secular variations of the earth's gravitational harmonic J_2 coefficient from Lageos and nontidal acceleration of earth rotation,' *Nature*, **303**(5920), 757-762, June 30, 1983.

PROBLEMS IN DATA COMPILATION

Heinrich Eichhorn and Carl S. Cole

Department of Astronomy, University of Florida
Gainesville, FL 32611
USA

ABSTRACT. Concepts involved in the estimation of target quantities and other adjustment parameters are critically discussd. We point out that one can find more accurate precepts for the reduction of data by utilizing all available constraints on all available data in the derivation of the reduction precepts. We introduce a measure for the **efficiency** of a set of adjustment parameters such that adjustments carried out using different precepts can be objectively compared. Finally, having applied our suggestions to a specific problem, we show that we have obtained estimates of a set of target quantities (in our case, star positions and proper motions) which have smaller formal errors than estimates of the same target quantities derived from the same input material but following traditional procedures.

1. INTRODUCTION

The compilation of a data set from observations of like data is one of the most frequent tasks of "data" reduction. While we shall in this paper emphasize the discussion of the compilation of complete catalogues of star positions (i.e., catalogues containing positions and proper motions), our considerations apply - mutatis mutandis - equally to the compilation of catalogues of magnitudes, radial velocities or generally any kind of precisely estimated like parameters from different sets of independent observations.

The situation is generally this: There is a set T of target quantities (e.g., the coordinates and proper motions of a certain set of stars), and there exist sets E_s (say, independent catalogues of observed positions) of estimates for subsets T_s of the quantities T. For each of the T_s, there is at least one other subset T_r whose intersection with T_s is not empty. This means that there are a number of series of observations of those quantities which make up the set T, such that each of the observation series contains the observation of at least one quantity which was observed in the course of at least one other series of observations.

The task, simply stated, is then to free the sets E_s of their "systematic errors" and to establish the compilation catalogue E, being a set of estimates of all elements of T. These estimates are computed by adjusting all available independent estimates of the elements of T after they have been freed from their

Celestial Mechanics **37** (1985) 263–275.

systematic errors. "Adjusting" is used here in the sense of estimating parameters which are connected to the directly observable quantities by a relationship whose analytical form is known. In the simplest case, this means taking the averages of the various available direct observations of a certain target quantity after the systematic errors have been removed. Generally, however, we have the case that quantities, such as, e.g., a position component α_0 at a certain epoch t^0 and the proper motion μ, are related to directly observable quantities. In our example these would be observed position components α_ν at an epoch t_ν which are related to α_0, t_0 and μ by a functional relationship, in our case $\alpha_\nu = \alpha_0 + \mu(t_\nu - t^0)$. The target quantities (α_0, μ) are then estimated by subjecting the pertinent observations (all α_ν) to an adjustment on the basis of the applicable analytical relationship.

After the "systematic errors" of the various E_s have been computed and removed from the observtions, the observations are regarded to be "on the same system". This sounds simple enough, and the various procedures have been canonized by many years of - successful - practice.

Yet, problems remain.

2. CONCEPTS OF DATA REDUCTION

2.1. The Concept and the Origin of the Systematic Error

Systematic errors in a set of observations would be called a "bias" by statisticians. The value of the bias depends on a number of parameters. Classical examples for systematic errors are, e.g., the personal equation and the magnitude equation which affect meridian observations, and color and magnitude effects which affect positions determined photographically.

It appears that the term "systematic error" is an unhappy misnomer. Investigating the origin of each such "systematic error in the observations" we discover invariably the following situation: What is published as a "set of observations" (one of our E_s) is not that at all, but rather a set of quantities which were calculated from the actually observed numbers on the basis of a certain model.

Unfortunately, modeling deficiencies are quite unavoidable. This is so for two reasons. The first is the practically always present uncertainty about the exact analytical form of the model, the second the unavoidable uncertainty about the values of the parameters in the reduction model. Remember that these parameters are never available as such, so that in their place, estimates obtained by some adjustment must be used.

2.2. The Analytical Form of the Reduction Model.

In calculating the quantities which are communicated as "observations", it is taken for granted that the reduction model, that is the form of the relationship between the actually observed quantities and those eventually communicated as observations is perfectly and completely known. Strictly speaking, this is an unrealistic assumption. While any successful accurate measuring is predicated on having at least a very good model for the analytical relationship between the **actual** measurements and the target quantities, the perfect relationship between these quantities is inaccessible. Consider, for example, photographic astrometry. For this purpose one may model the general features of the relationship between the rectangular coordinates of the stars' images on the plate (exposed in the focal "plane" of the objective) and their positions on the sphere in terms of the equations which describe an ideal pinhole (i.e., gnomonic) projection. As soon as one is interested in an accurate model, one runs into a long series of obstacles. No real objective, even if it consisted of components which were all perfectly shaped to specification (a technological impossibility) produces an ideal gnomonic projection in a focal plane. Many books have been written on the various aberrations which depend on just about every parameter that characterizes the image and the star which produces it. Since these aberrations produce in general small deviations from gnomonic projection, they are modeled by first (and in rare case higher) order approximations. But even if we had a perfect model for the pertinent imaging characteristics of the ideally shaped optics, we would know nothing about those aberrations which are introduced because the optical components can never be manufactured to perfect shape. These deviations from perfect shape might be treated as irregular, and therefore accidental, were it not for the uncomfortable fact that they will - ceteris paribus - manifest themselves in the same way on all plates taken with the same optics. One might gain control of them, but evidently only at the cost of introducing additional reduction parameters, an undesirable procedure; cf. Eichhorn and Williams (1963). Even more difficult to control are the irregularities of the plates and generally, the unavoidable failure of the photographic film to coincide ideally with the focal plane or -generally - focal surface. Reality is always more complicated than any model.

It would therefore be best to avoid the misleading and vague term "systematic errors" altogether and instead speak of the **effects of modeling deficiencies.** This term suggests that something may be undertaken to correct these deficiencies, while the term error - perhaps from the frequent association with "accidental" - connotes that one is dealing with something unavoidable.

2.3 The Accuracy of the Parameters in the Reduction Model

Any model which purports to furnish a relationship between actual observations and certain target quantities contains adjustment parameters which are not known a priori but must be estimated in the course of the adjustment of the observations. The parameter estimates driven by a certain postulate, e.g., "least squares," maximum likelihood or some such, differ of course from the unobtainable actual values of the parameters themselves. The parameter **estimates** are all one has for use in the reduction model. Even if this were a completely accurate representation of the geometry and physics of the situation which is thereby described, the values of the target quantities of the model, computed by use of the parameter estimates, would differ systematically from those values which one would obtain if the actual values of the parameters were available. This phenomenon, for which the name "parameter variance" has been introduced, means that there cannot be a perfect reduction of data to a "system" (Eichhorn and Williams 1963). In particular, the following is important to note: If the adjustment of the observations, involving the estimation of the model parameters was performed by the least squares principle which minimizes a certain quadratic form (cf. Brown 1955), the **value of the same quadratic form** (in the simplest case the sum of the squares of the adjustment residuals) **would inevitably increase** if the parameter estimates in the model were replaced by the true values of the parameters, which means that the true values would inevitably produce a worse "fit" than the estimates!

3. CRITIQUE OF THE TRADITIONAL PROCEDURES

3.1. Systems and their Relationships

It has been pointed out before (Eichhorn 1982), that a system must be considered a set of discrete estimates of certain target quantities. The FK4, for example, is a set of estimates of positions and proper motions of stars at a certain epoch with respect to a specified coordinate system (the Q system). The long-term relationship of the Q system to an inertial system is specified by certain adopted values for the constants of precession, the motion of the equinox and the rate of the obliquity; the short-term relationship to an inertial system is specified by certain adopted parameters in nutation theory. The system of the FK4 (or any fundamental catalogue) can therefore not be regarded as an inertial system by definition (it may, in fact, not even be, or rather it is now known not to be, self-consistent) but only as an estimate of one. After all, the (unavoidable accidental) errors in the data which constitute a fundamental system render the system patently (albeit only ever so slightly) nonself-consistent.

While no King's Way has yet been delineated in which one would reduce one catalogue to the system of another without ambiguity or leeway, various methods have been used in the past. The simplest took the straight averages of the deviations (system minus new catalogue) of all data in certain domains and regarded these as the values of the correction function to be applied to reduce the new data to the system at the center of the domain. The emergence of computers has made possible the establishment of analytical interpolation models. The numerical parameters in the model are the result of some objective procedure (Schwan 1977). Not surprisingly, in spite of the fundamental difference in the procedures for establishing correction functions, the tabular and analytical models yield similar results. (Bien et. al. 1978).

3.2. Critique of the Theoretical Foundations of the Tabular Method

In spite of satisfactory numerical results, the tabular method itself is conceptually poorly defined and lacks precision. Tablular methods have no sound theoretical justification other than that they have performed adequately for the centuries when computers were not available.

In the simplest case of the nonanalytic tabular method, the actual correction (not to be confused with its estimate) to be applied to the positions given in a catalogue to reduce them to the system of another catalogue is considered a fixed constant in each subjectively defined domain. The model for the observed error of, e.g., the position of star μ in domain ν $\delta_{\mu\nu}$ is:

$$\delta_{\mu\nu} = p + \varepsilon_{\mu\nu} , \qquad (1)$$

where p_ν is the fixed but unknown systematic error (negative system correction) in domain number ν and $\varepsilon_{\mu\nu}$ is a random error, distributed with mean zero and variance $\sigma_\nu{}^2$. Under these assumptions one invokes the principle of least squares and minimizes the sum of the squares of all $\varepsilon_{\mu\nu}$. In this way one obtains an estimate p_ν^* for p_ν as the mean of all $\delta_{\mu\nu}$.

$$p_\nu^* = \frac{1}{n_\nu} \sum_{\nu=1}^{n} \delta_{\mu\nu} \quad ; \qquad (2)$$

n_ν, of course, being the number of differences formed in the νth domain. An unbiased estimate, S_ν^2, for σ_ν^2 is:

$$S_\nu^2 = \frac{1}{n_\nu - 1}\left(\sum_{\mu=1}^{\nu} \delta_{\mu\nu}^2 - n_\nu p_\nu^{*2}\right)$$

So far, this is statistically sound, because the system corrections are uncorrelated, under the given assumptions. The estimate of their (diagonal) covariance matrix is:

$$S^2 = \mathrm{diag}\,(S_1^2, S_2^2, \ldots, S_\mu^2),$$

where m is the number of domains which are involved in the process.

Unfortunately, the model (1) proves inadequate. Note that the domains cannot at the same time be chosen small enough to model the fine structure of the systematic differences and yet large enough not to mask the random errors of the observations. Usage, handed down from one generation of data manipulators to the next, calls for choosing smaller domains and then "smoothing" each estimate, p_ν^*, with its immediate neighbors. It is during this process of smoothing that the underlying assumptions are invalidated, thus causing the theoretical foundations to lose their applicability to the specific instant. However, this does not mean that some kind of numerical results will not be produced.

If smoothing is involved, the model for the observed error of the position of star μ in domain ν' is:

$$\delta_{\mu\nu'} = \sum_{\nu=1}^{m} a_\nu p_\nu + \varepsilon_{\mu\nu'} \quad , \qquad (3)$$

where the a_ν are subjectively chosen smoothing constants and the smoothing occurs over domains in the neighborhood of domain ν'. The method of least squares would yield the estimates, p_ν^* by minimizing the quantity

$$\sum_{\nu'=1}^{m} \sum_{\mu=1}^{n_{\nu'}} \varepsilon_{\mu\nu'}^2 \; .$$

This, however, is not done in conventional practice; The p_ν^* are rather obtained from equation (2).

The presumptions of the model function and those of the procedure for estimation the model parameters thus contradict each other. The model function (3) is predicated on the assumption that the p_ν are correlated, thus giving justification for the smoothing process, while the estimation of the model parameters from (2) is based on the assumption that the p are independent.

Furthermore, and independent of the criticism above, estimation of the covariance matrix of the model parameters is made computationally intractable in all but the simplest of the tabular methods, thus these matrixes have rarely, if ever, been calculated. Without an estimate of the covariance matrix of the model parameters, a tabular method cannot be objectively compared to other methods. The only "test" of a tabular method is whether or not it produces a satisfactory result. This is almost by necessity a subjective judgement.

The literature is full of diagrams that plot tables of empirically derived functional values, at discrete arguments, together with their "error bars" and regards as the definitive curve which represents the model function the polygonal arc which connects the derived estimates by straight lines. The authors know better than that, of course. If the "error bars" indicate standard errors, we know from probability theory that the true curve will intersect only about two thirds of the error bars and almost never go through one of the points. Rather than drawing a wild zig-zag line which passes through all error points, it would be more reasonable to postulate a smoother model function, and draw a straighter curve such that only about two thirds of the error bars are intersected. Everything else patently constitutes overmodeling. If, for example, a certain model function (one of the correction functions to the FK4 system, say) has been empirically derived independently by two different procedures and if one manages to draw a curve through the combined available data points which intersects about two thirds of both sets of errror bars, one is obliged to consider the two determinations as free from contradictions, even if they give the appearance of showing systematic differences.

3.3. The Form of the Model Functions

The model function depends in any case on the position of the star in the Q-system and almost certainly also on the magnitude and likely also on the color equivalent of the star to whose position it is to be applied. Unfortunately, the range of at least the magnitudes of the stars in the FK4 does not reach as faint as that of most independent catalogues that are to be "reduced to the FK4 system". Unless one knows the magnitude dependent part of the model function well enough to extrapolate to fainter magnitudes from the limited magnitude interval which the fundamental catalogue and the independent catalogue have in common, one cannot really meaningfully state that the fainter stars in the independent catalogue have also been put on the FK4 system; strictly speaking, the FK4 system is not even defined for these faint magnitudes. The only way to avoid this dilemma is to attempt to produce catalogues which simply have no magnitude equation. (Not that this has not been tried!)

3.4. The Importance of Using All Available Constaints.

As a typical task catalogue astronomy, each one in a group of numerous catalogues (independent catalogues) is individually and independently reduced to the system of a less numerous catalogue (a fundamental catalogue, say). By "individually and independently" we mean that each catalogue in the group is "reduced to the system" of the fundamental catalogue by a procedure which involves only the data in the fundamental catalogue and the catalogue being then reduced, but not data in any of the other catalogues in the group. The thusly reduced data from the independent catalogues are subsequently combined (averaged) into a compilation catalogue "on the system" of the original catalogue. Well entrenched though this process may be, and no matter how well it may in the past have served in practice for the production of numbers, every element of the process outlined above is subject to vagueness, ambiguity and lack of precision of the concepts it implies.

Consider that the accuracy with which a correction function can be empirically established, be it as a formula (analytically) or in the form of a correction table, grows almost monotonically with the number of points available for which like data can be compared. More precisely, this means the following: Given a model function with a certain number of free parameters, the variances of the parameter estimates will - ceteris paribus - decrease (almost) inversely with the increase in the number of comparison points.

We have pointed out above (Sec. 2.3) that the establishment of the reduction model **always** involves the estimation of certain model parameters whose actual values never become available, and that therefore the differences between the parameter estimates used and the true values of these parameters **must** produce significant systematic differences in the reduced positions of the non-FK4 stars in the two catalogues as a consequence of parameter variance. This means, of course, that the unavoidable imperfections of the establishment of empirical model functions become obvious when those position comparisons are extended to stars which could not be utilized in the comparison which were the basis for the establishment of the original reduction models. Obviously, the "best" available reduction models for both catalogues have therefore not yet been obtained, they would (almost) have to eliminate any systematic differences between the positions of the field (i.e., non-FK4) stars after the reduction. If this is accomplished, it will (cf. Sec. 2.3) unavoidably increase the dispersion of the residuals (FK4 position minus independent position) of the FK4 stars after the "definitive" reduction, because the preliminary reduction was, after all, based on minimizing this very dispersion or at least its equivalent.

Fears, however, that the incorporation of the non-FK4 stars will interfere with the proper duplication of the FK4 system are inappropriate; remember that after all, the parameters of the original reduction model, based on comparison to FK4 data only, are merely estimates and that the addition of independent new data can **never** decrease the accuracy of an estimate. Remember that the estimation of the reduction model is predicated on the minimization of the variance of the adjustment residuals. Furthermore, the square of the average of the dispersions between the values calculated from the true and the estimated model functions, respectively, is exactly the parameter variance, which can always be computed as long as the covariance matrix of the estimates is known. Therefore, since variances can be added, one should expect that the variance of the target quantities obtained by using the true values of the reduction parameters exceeds the variance of the target quantities based on the estimated model parameters exactly by the parameter variance. **Only if a subsequent reduction produces a variance which exceeds that produced by the original solution by an amount which is significantly larger than the parameter variance need one fear that the new solution found with the use of additional material does**

not represent the target system as well as the original solution.

Also, one must expect the square of the dispersion of the average systematic differences between the positions of identical non-FK4 stars from two different catalogues to equal the sum of the parameter variances of those two estimated model functions that were used to reduce these catalogues to the reference system. If the model functions are adjusted to minimize, **simultaneously with the minimization of the variance of the adjustment residuals of the independent catalogue against the reference system, also the variance of the differences between the positions common to only the independent catalogues,** one will likely obtain better, that is, more accurate, reduction models than by comparison of the independent catalogues with the reference catalogue only.

As we have already stated, it is to be expected that such models will increase the variances of the adjustment residuals against only the reference catalogue, and as long as this increase is not by more than the parameter variance of the original estimate, there is nothing wrong. If, however, the incorporation of additional, e.g., inter-catalogue, comparisons increases the original variance of the adjustment residuals against the reference catalogue by an amount which is significantly larger than the parameter variance of the originally determined reduction model, one must suspect that the analytical (or numerical) form of the adjustment model was inadequate. This would happen, e.g., if the reduction models for catalogues that have a significant magnitude equation had been derived without allowing for magnitude dependent effects. One would then see the following: The independent catalogues would be reduced, basically to their mean, at the most frequent magnitude, typically fainter than the most frequent magnitude of the reference catalogue. The failure to provide in the model for a really existing magnitude equation would make it impossible to fit the catalogues satisfactorily to each other and simultaneously to the reference catalogue, and this would, of course, be evident in the residuals of the reference catalogue stars when the independent catalogues are compared to the reference catalogue **after** a reduction model which does not accommodate magnitude effects was determined which took into account the fitting of the independent catalogues to each other.

4. TOWARD CRITERIA FOR JUDGING ACCURACY AND VALIDITY OF REDUCTIONS

4.1. The Number of Independent Parameters in the Tabular Method

Based on Brosche's (1966) ideas, Schwan (1977) devised a highly sophisticated analytical procedure for deriving correction functions to catalogues. One hears occasional criticisms of this analytical method which states essentially that the analytical approach is not capable of representing properly the "fine structure" of the various correction curves, or at least not with a "reasonable" number of parameters. Before such criticism is agreed to, one might ask two questions: The first is whether the "fine structure" which was found by the tabular method is at all real. If it is based on a correction curve which goes through all data points (without missing any error bars) it is safe to say that the fine structure is probably spurious and should not have been represented at all. (See Sec. 3.2 above). The second question concerns the equivalent number of independent parameters in the tabular determination of a model function. If this is done by taking the means of the sample values in certain domains, the equivalent number of independent parameters is obviously equal to the number of domains. If one derives, for example, systematic corrections of a catalogue between 30° and 60° to the FK4 system in regions $40^m \times 10°$ (so that one would typically, have three FK4 star per box) one would have 108 boxes and thus 108 independent parameters. The raw values are of course subjected to some smoothing and interpolation procedure before they are regarded as representative estimates of the function which they are purportedy sampling; this cuts down the efficiency of independent parameters because each value depends in part on the data which were used to derive the values of its immediate neighbors. It appears plausible that the taking of simple running means would cut the efficiency of independent parameters by a factor of 3. If, however, the running means were to be taken by assigning reduced weights for the neighboring values, the reduction of the efficiency number of free parameters would be by a factor less than 3. In this case, if straight running means were taken in both right ascension and declination (ignoring the difficulties into which one runs in declination with this at the borders of the zone), the efficiency would have been cut by a factor of 9.

4.2. A Measure of the Efficiency of a Set of Model Parameters.

This opens an interesting question: Obviously, the information carried by a vector **x** of correlated estimates (whose covariance matrix **Q** is known) is partly redundant because of the non-vanishing correlations between the estimates. The question is : What is the information content carried by these estimates? Remembering that the matrix $\mathbf{P}^T\mathbf{QP}$ is diagonal if **P** is the (orthogonal) matrix of the normalized eigenvectors of **Q** and that it is also the covariance matrix of the (therefore noncorrelated) linear transforms $\mathbf{y} = \mathbf{Px}$ of **x**, we might regard the number $e = |\mathbf{Q}| / q_{11} \cdot \ldots \cdot q_{nn}$, that is, the product of the variances of the components of **y** divided by the product of the variances of the components of **x** as a measure for the **efficiency** of the information carried by the estimates **x**. For any set of noncorrelated estimates we would have $e=1$ which is evidently the largest value this number may assume. (Brosche's method and Schwan's generalization calculate uncorrelated parameters in the model function and thus automatically assure $e= 1$.) In comparing the accuracy of the fit between an analytically estimated model function, and one found in the form of a table one would first have to establish that they are comparable, that is that the number of numerically derived table entries, multiplied by the appropriate e, and the number of parameters estimated in an analytical model, multiplied by the number e in this case, have the same value. With this verified, that model is obviously superior which produces the smaller dispersion of the residuals.

5. RESULTS: THE COMPILATION OF A COMPREHENSIVE CATALOGUE.

Corbin (1977) has compiled a catalogue (NIRS: Northern International Reference Stars) on the system of the FK4 by reducing 64 independent catalogues, with mean epochs ranging from 1889 to 1959 to the FK4 system and then constructing from the reduced data a complete catalogue of stars between -5° and +90°. Corbin found his reduction models in steps: he first constructed an intermediate, "base" catalogue by reducing to the FK4 those ten of the independent catalogues which had a sufficient number of stars in common with the FK4, and subsequently regarded the much more numerous base system as essentially an extension of the FK4 system. The base catalogue then served as the reference catalogue for the compilation of the final catalogue, in the sense that Corbin assumed that the reduction models which reduce those independent catalogues which had not contributed to the establishment of the base catalogue were at the same time valid precepts for the reduction of these catalogues to the FK4 system.

One can see that Corbin's cantilevering procedure is a first step toward what we are advocating here, because the reduction precepts of most of the independent catalogues were derived not from direct comparison with the original reference system - the FK4 - but from comparison with a more numerous intermediary system, the base catalogue. For further details we refer to Corbin's most careful and painstaking investigation.

In preparing a re-compilation of Corbin's NIRS, from exactly the same material which he used, we recompiled the base catalogue, on the system of the FK4, from the same catalogues which Corbin used for this purpose, except that we derived our reduction models not only from comparison with the FK4, but also utilized the intercomparison of the various catalogues. (See Sec. 3.4 above)

One could , of course, set up the problem for solution in closed form, solving simultaneously in a block adjustment for all parameters in the reduction model. This is completely analogous to the so-called overlap solution in constructing a photographic star catalogue. It has been shown that an iterative process works and converges toward the same solution as the direct, closed approach, but under the circumstances with less of a computing effort and without the requirement that large matrices be inverted.

As first step we essentially repeated Corbin's steps which led to the establishment of his base catalogue: After comparing the independent catalogues with the FK4 (at the epoch of the catalogue positions, of course) in sectors of one hour width in right ascension and zones one degree wide in declination, we constructed $\Delta\alpha_\alpha$, $\Delta\alpha_\delta$, $\Delta\delta_\alpha$ and $\Delta\delta_\delta$, which were smoothed using the same coefficients as Corbin. These corrections were next applied to the positions in the independent catalogues, and used for the derivation of the positions and proper motions of **our** first iteration base catalogue. Had we traced Corbins's steps exactly, this would have duplicated his base catalogue and the standard errors of the data there derived by him.

We next started a process of iterations: The positions and proper motions in the last available version of our base catalogues were used as the reference catalogue for the derivation of reduction models of the individual independent catalogues to the system, ostensibly that of the FK4, by essentially the same procedure which produced the first approximation to our base catalogue; note, however, that in the course of any but the first iteration, there is a reference position available for each star in each of the independent catalogues involved. The iterations were stopped after the results from two successive ones yielded essentially identical results. In our case, this happened after the fifth iteration.

Table I lists below the like data in Corbin's catalogue, the average formal standard errors of our base catalogue positions after each iteration.

TABLE 1

average rms errors

	$\sigma_\alpha \cos\delta$	σ_δ	$\sigma_\mu \cos\delta$	σ_{μ_δ}
	(unit: 0."001)		(unit: 0."01/century)	
Corbin's Base Catalogue	64	77	37	40
1st iteration	65	77	39	40
2nd iteration	62	75	37	40
3rd iteration	61	75	36	40
4th iteration	61	74	36	40
5th iteration	60	74	36	40

Table I illustrates well the improvement - at least as judged by the formal dispersion of the data achieved by the iterations - which we achieved by employing the intercomparisons of the independent catalogues in additon to comparisons of the independent catalogues with the reference catalogue (i.e., the FK4) for the estimations of the reduction model.

The diagrams (Figs. 1 through 5) show the $\Delta\alpha_\delta \cos\delta$, averaged over five degree zones for the catalogue W20, one of those which was incorporated into Corbin's (and our) base catalogue.

Figure 1 is a plot of the differences $\Delta\alpha_\delta \cos\delta$ in the sense FK4 - W20, in units of 0."001, against the declinations. Figs. 2 and 4 show this plot, for the differences FK4 - W20 and base - W20, respectively, after W20 was a reduced to the FK4 system by the traditionally definitive procedure.

Were this the definitive reduction, both Figs. 2 and 3 should show points scattered around zero, with the dispersions of the scatter basically determined by the average rms error of the positions in W20.

Observe, however, that

a) neither Fig. 2 nor Fig. 3 show points scattered around zero without significant trend - this shows that the reduction derived in the step that led to Fig. 2 is not definitive, and

b) that Figs. 2 and 3, while quite similar, are not identical - this shows that the first approximation of the base catalogue - the end product of the canonical classical procedure - while quite close to that of the FK4, is by no means indistinguishable from it.

Figures 4 and 5 represent what we regard as the final result. While the points in Fig. 4 all indicate positive values (average : 47 $\pm$ 26 s.e.) - showing that a magnitude equation would have been necessary to achieve a good reduction of W20 to the FK4 system, the points lie basically on a straight line, which indicates that the declination dependent part of the reduction model is well described. The same is true of the corresponding Fig. 5, although there, the points (average: -10 $\pm$ 18 s.e.) lie rather below the line $\Delta\alpha$=0., they also show less scatter, which indicates that the fit of the W20 to the system defined by our base catalogue, though not systematically different from that defined by the FK4, is still tighter, which is what one can expect. From the diagrams one cannot judge whether the magnitude equation is in the W20 or in our (definitive) base system, although one is naturally inclined to suspect the individual catalogue.

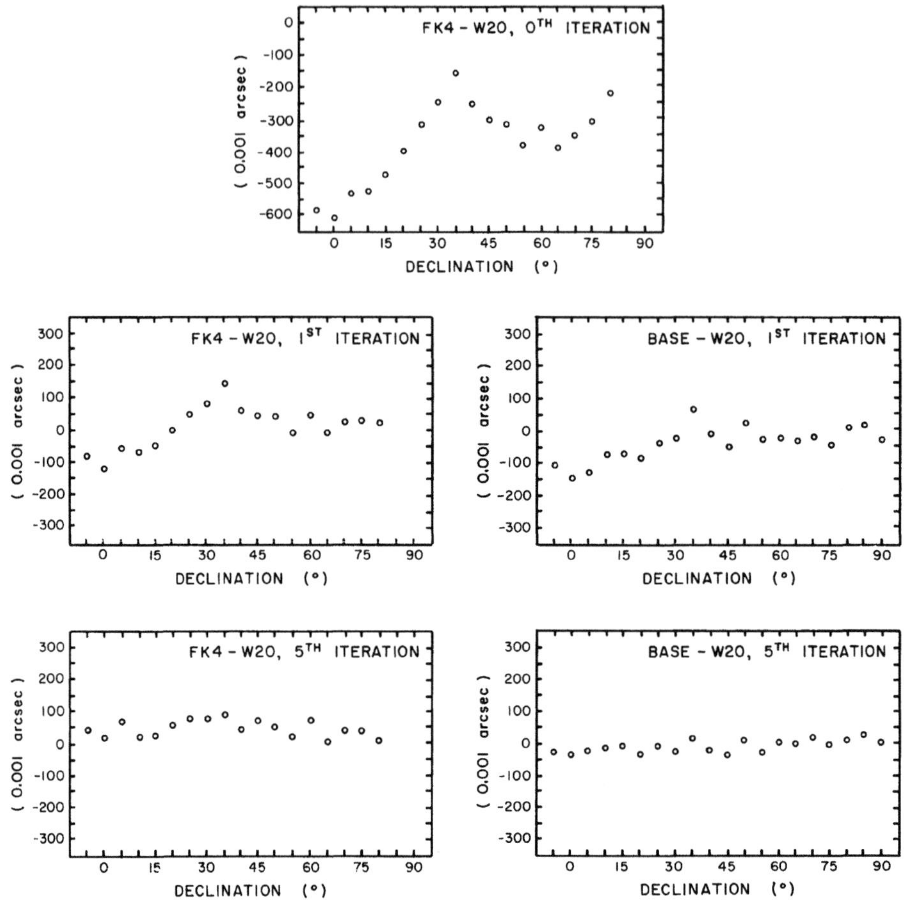

Figures 1-5. Systematic differences in right ascension averaged over 5^0 zones of declination ($\Delta\alpha_\delta \cos\delta$).

Finally, a comparison of Figs. 2 and 4 shows that the W20 is indeed brought closer to the system of the FK4 by involving all the positions in the W20 to calculate the model for reducing this catalogue to the FK4 system.

ACKNOWLEDGEMENTS

The authors would like to thank J. Hughes and T. Corbin of the U.S. Naval Observatory and W. Warren of the Astronomical Data Center for provided data used in this investigation. G. Contopoulos, who critically read a preliminary version of the manuscript, made several cogent suggestions. One of us (CSC) performed his part of the work reported here as beneficiary of a Graduate Research Assistantship from the Division of Sponsored Research of the University of Florida.

REFERENCES

Bien R., Fricke, W., Lederle, T., and Schwan, H. 1978. Veröff. Astron. Rechen-Institut Heidelberg No. 29.
Brosche, P. 1966. Veröff. Astron. Rechen - Institut Heidelberg No. 17.
Brown, D. C. 1955. Ballistic Research Laboratories Rept. No. 937. Aberdeen Proving Grounds, Maryland.
Corbin, T. 1977. The Proper Motion System of the AGK3R. University Microfilms, Ann Arbor. (Dissertation, University of Virginia)
Eichhorn, H. 1982. in: Automated Data Retrieval in Astronomy, p. 103. C. Jaschek and W. Heintz (eds.). Dordrecht.
Eichhorn, H. and Williams, C. A. 1963. Astron. Journ. **68,** 221.
Schwan, H. 1977. Veröff. Astron. Rechen - Institut Heidelberg No. 27.

THE SOUTHERN REFERENCE STAR PROGRAM: A 20-YEAR INTERNATIONAL PROJECT NEARING COMPLETION

Clayton Smith and Edward Jackson
U.S. Naval Observatory
Washington, D.C.

1. INTRODUCTION

In 1961, 12 of the transit circle telescopes of the world in both the northern and southern hemispheres started an international program of observation of about 20,500 stars within the declination limits +5 to -90 degs. These stars are to be used as the reference stars in the reduction of the southern hemisphere second Cape Photographic Catalog (CPC2) (see Nicholson, et al. 1984) and are known as Southern Reference Stars(SRS). They are the southern hemisphere counterparts of the AGK3R, the reference stars for the AGK3 photographic catalog. In the zone of overlap between the SRS and AGK3R lists from −5 to +5 deg. declination, the SRS and AGK3R lists are identical. The entire ensemble of reference stars including both the SRS and AGK3R lists is frequently referred to as the International Reference Star (IRS) list. Most of the stars are in the visual magnitude range from about 6.5 to 9.5 and the stars have been rather carefully chosen for their good observational histories to ensure the availability of a sufficiently large body of data to calculate accurate proper motions. Another criterion for selection was to achieve as nearly as possible a uniform distribution of the stars on the celestial sphere with a density of about 1 star per square degree. For a summary of the work on the compilation of the AGK3R see Smith (1980) and for a discussion of the work on the formation of the proper motion system see Corbin (1974,1978). The present paper describes progress in the compilation of observations of the SRS program which were made from 1961 to 1973 into a reference catalog of mean positions at the mean epoch of observation referred to the system of the FK4 catalog.

2. THE DATA BASE

Among the 12 transit circle telescopes which participated in the SRS observing effort, seven were located in the northern hemisphere at Abbadia(Hendaye, France: N43° 50'), Bordeaux(Floirac, France: N44°50'), Bucharest(Romania: N44° 25'), Nicolaiev(U.S.S.R.: N46°58'), San Fernando (Spain: N36° 28'), Tokyo(Japan: N35° 40'), and Washington(U.S.A: N38°55'). Five transit circles were located in the southern hemisphere at Perth (Bickley, Western Australia: S32° 1', Bergedorf instrument), the Cape of Good Hope(Capetown, South Africa: S33° 56'), San Juan(Argentina: S31°31'),

Celestial Mechanics **37** (1985) 277–284.

TABLE 1

SOUTHERN REFERENCE STARS (SRS)

DATA BASE FOR THE COMPILED CATALOG OF POSITIONS (1961 - 1973)

CONTRIBUTOR	SRS DECL. ZONE	SRS-# OBSNS (# STARS)	FK4-# OBSNS (#STARS)
ABBADIA	+5 TO -15	6170 (1550)	2750 (350)
BORDEAUX	+5 TO -15	7323 (1620)	3420 (350)
BUCHAREST	+5 TO -10	5070 (1170)	3200 (330)
CAPE	-30 TO -64	32767 (8160)	3740 (370)
LEONCITO (USNO 7″)	+5 TO -90	65960 (20220)	51410 (890)
NICOLAIEV	0 TO -20	18090 (5950)	8900 (410)
PERTH (BERGEDORF)	+5 TO -90	70750 (20150)	17000 (860)
SAN FERNANDO	-10 TO -30	15410 (3650)	7280 (470)
SAN JUAN	-40 TO -90	22820 (7180)	10530 (340)
SANTIAGO	-25 TO -47	28780 (5990)	8090 (450)
SANTIAGO (PULKOVO)	-47 TO -90	25230 (5480)	9320 (280)
TOKYO	-10 TO -30	16320 (3630)	10510 (540)
WASHINGTON (USNO 6″)	+5 TO -30	22210 (9830)	12040 (500)
		336900 (20495)	147920 (890)

TOTAL NUMBER OF FK4 AND SRS OBSERVATIONS = 484820

AVERAGE NUMBER OF OBSERVATIONS PER SRS = 16

AVERAGE NUMBER OF OBSERVATIONS PER FK4 = 166

Santiago(Chile: S33°25'), and El Leoncito(Argentina: S31°48', Washington Seven-inch transit circle). Most of the observations were made visually, but one of the instruments (Perth) had a photo-electric detector.

Substantial numbers of observations of stars from the FK4 catalog were interspersed among observations of the SRS stars in order to facilitate the reduction of SRS positions to the FK4 system. Table 1 shows the approximate number of observations of SRS and FK4 stars and the declination limits of the zone of observation for each participant. It is evident from Table 1 that about one-third of the total number of observations were of stars from the FK4 catalog. This high proportion of observations of FK4 stars helped to assure that the contribution of each participant in the SRS program would be capable of a very accurate reduction to the system of FK4 by direct observational means. The average ratio, by observatory, ranged from a high of about eight FK4 per 10 SRS observations to a low of about one FK4 per 10 SRS observations. The optimum or target value adopted by most observers was about two to three FK4 per 10 SRS. Overall, the total number of SRS plus FK4 observations is about 481,500. The average number of observations per SRS is 16 and per FK4 is 166. This may be compared with the statistics of the AGK3R observing campaign (1956-1963) in which each AGK3R star was observed about 10 times and each FK4 star in the AGK3R zone was observed about 110 times.

The coverage by a large number of transit circles in the zone from +5 to -30 degrees in declination is excellent, as shown in Table 1. Parts of that zone were observed by as many as eight participants while in the sub-zone from -20 to -25 there were only five participants. In the region south of declination -30 degrees, there are many zones where only four participants contributed. In the last 25 degrees from -65 degrees to the south pole four instruments participated, with important series of observations made at San Juan, Santiago, Perth and Leoncito. Even though every effort has been made to get as close to the FK4 system as possible by using special observing techniques and appropriate reduction procedures, the small number of contributors in the zone between -65 deg and the south pole make it likely that there the systematic quality of the SRS catalog will be relatively lower than elsewhere. The poor quality of the FK4 system itself at declinations south of -30 deg (Lederle 1978) will add to this problem.

The data base was not assembled in its entirety until late in 1984, about eleven years after the last observations had been made. This may be taken as an indication of the difficulty of adequately discussing large bodies of observational material with complex internal relationships when the best transformation to a particular catalog system is desired.

3. THE PRELIMINARY COMPILED CATALOG

It was our initial presumption that all participants had brought their own observations of the SRS to the system of FK4. On the basis of that presumption, a preliminary version of the SRS catalog was compiled in which the results of each participant in the form of mean observed minus computed apparent places, (O-C)'s, in right ascension and declination referred to the mean epoch of observation were combined for a given SRS with equal weight, regardless of the number of observations.

Differences, (I-CM), between the individual instrumental mean (O-C)'s, I, and the compiled mean (O-C)'s, CM, were formed for each star in right ascension and declination and collected in tabular form by instrument and by five degree zones in declination. A comparison of the (I-CM) differences showed a significant lack of agreement among the instrumental systems, even though all had been referred to the system of FK4 by observational means. The systematic differences were strongest in right ascension in the region from -50 degrees to the south pole where a range of (I-CM) values of the order of 0.025 s among the instruments was found. In declination, the range of (I-CM) values was uniform at 0.30 arcsecs over the entire SRS region from +5 to -90 degrees declination. On the basis of this result, it was clear that further study of the instrumental systems was necessary in order to improve the systematic reduction to the FK4 system.

4. IMPROVED REDUCTION TO THE SYSTEM OF FK4

Most of the SRS participants reported observations of FK4 stars which were made concurrently with the SRS program. We used the reported (O-C)'s of the FK4 stars to form tables of systematic differences in which the (O-C)'s were binned into cells one hour by five degrees in right ascension and declination, respectively. From these tables of systematic differences, interpolated and smoothed (O-C)'s were obtained for application to the (O-C)'s of the SRS stars. It was found that interpolation by zones of declination alone did not adequately refer SRS observations to the FK4 system, and that it is necessary to interpolate in right ascension, also. With the exception of Nicolaiev and San Juan, smoothed right ascension- and declination-dependent differences were applied to all SRS observations in order to improve the reduction to the FK4 system. From Nicolaiev, we received only results of SRS observations, so we had no possibility of making a new reduction to the FK4 system. In the case of San Juan, it was clear that a detailed reduction to the FK4 system had already been made, so no further improvement was possible. On this improved basis, a new compilation was made for mean (O-C)'s of the SRS.

5. THE REVISED COMPILATION

The improved (O-C)'s for a given SRS were combined with a weight equal to the number of observations reported to us by the observatory up to a maximum weight of nine. This cap on the weight was necessary to prevent the weight of dozens of observations of certain SRS by a few participants from dominating the averages in a program where two to four observations per star was typical. Substantially better agreement was clearly evident when the revised values of (I-CM) were intercompared. For example, in right ascension, the largest systematic differences dropped from the 0.025 s range to about 0.017 s in the declination zone from -55 to -90 degrees. In the region of best agreement, from +5 to -25 degrees, the largest systematic differences are of the order of 0.006 s. We presume

TABLE 2

DISTRIBUTION OF THE FREQUENCY OF THE MEAN ERRORS OF THE MEAN POSITIONS BY MEAN ERROR

MEAN ERROR	FREQUENCY R.A.	FREQUENCY DECL.	MEAN ERROR	FREQUENCY R.A.	FREQUENCY DECL.
0"01	103	51	0"26	7	50
0.02	505	136	0.27	3	33
0.03	1082	355	0.28	2	28
0.04	1438	627	0.29	-	16
0.05	2284	976	0.30	2	13
0.06	2552	1266	0.31	1	12
0.07	2248	1632	0.32	2	4
0.08	2284	1761	0.33	-	5
0.09	2016	1790	0.34	-	5
0.10	1198	1840	0.35	-	1
0.11	1148	1696	0.36	-	2
0.12	828	1612	0.37	-	2
0.13	502	1396	0.38	-	1
0.14	388	1126	0.39	-	-
0.15	249	976	0.40+	16	13
0.16	155	749			
0.17	108	600			
0.18	71	462			
0.19	41	351			
0.20	32	272			
0.21	19	207			
0.22	11	142			
0.23	9	110			
0.24	2	84			
0.25	3	59			

AVERAGE MEAN ERROR: 0"08 0"11

50% OF THE M.E.'S IN R.A. AND DECL. ARE LESS THAN 0"07 AND 0"10, RESPECTIVELY.

95% OF THE M.E.'S IN R.A. AND DECL. ARE LESS THAN 0"14 AND 0"20, RESPECTIVELY.

TABLE 3

DISTRIBUTION OF THE MEAN ERRORS OF THE MEAN POSITIONS BY DECLINATION ZONE

DECL. ZONE	AVERAGE MEAN ERROR R.A.	AVERAGE MEAN ERROR DECL.	DECL. ZONE	AVERAGE MEAN ERROR R.A.	AVERAGE MEAN ERROR DECL.
+5 TO 0	0"08	0"11	-45 TO -50	0"06	0"09
0 -5	0.08	0.11	-50 -55	0.07	0.10
-5 -10	0.08	0.11	-55 -60	0.07	0.10
-10 -15	0.08	0.12	-60 -65	0.08	0.10
-15 -20	0.09	0.13	-65 -70	0.07	0.12
-20 -25	0.08	0.11	-70 -75	0.07	0.12
-25 -30	0.08	0.11	-75 -80	0.08	0.11
-30 -35	0.08	0.11	-80 -85	0.09	0.11
-35 -40	0.07	0.10	-85 -90	0.08	0.12
-40 -45	0.06	0.09			

that the reason for the excellent systematic agreement in this zone comes about because of the excellent systematic quality of the FK4 catalog itself in that zone, just as the poorest agreement among the participants is found, not unexpectedly, in the zone where the systematic quality of the FK4 catalog is worst. In declination, the systematic agreement improved from the 0.30 arcsec level to about 0.20 arcsec over the entire SRS region from +5 degrees declination to -90 degrees.

This step is viewed as the first iteration in an objective approach to the problem of developing correct weights for each contributing observatory.

6. ANALYSIS OF THE RESULTS OF THE REVISED COMPILATION

The frequency distribution of the mean errors of the mean positions is given in Table 2, where the counts are binned in units of 0.01 arcsec. In right ascension, the distribution peaks at 0.06 arcsec, and in declination at 0.10 arcsec. The average mean error is 0.08 arcsec in right ascension and 0.11 arcsec in declination. A clear picture of the relatively narrow shape of the distribution is demonstrated by the fact that 95% of the SRS with more than three observations have mean errors less than 0.14 arcsec in right ascension, and 0.20 arcsec in declination.

The distribution of the mean errors of the mean positions by declination zone is shown in Table 3. There it can be seen that the mean errors are essentially equal over the entire southern hemisphere. At the northern and southern extremes, there is a small increase of 0.02 to 0.03 arcsec in both right ascension and declination relative to the mid-declinations. This is most probably caused by the larger zenith distance at which the SRS in the extreme zones are observed from mid-latitude observatories in both the southern and northern hemispheres.

An objective technique first suggested by F.P. Scott (1967) was used to compute weights of the various contributors. The variances of the (I-CM) values are used as an estimator of the quality of the work of a particular contributor. From the variances, weights proportional to the reciprocal of the variances were calculated, and are shown in Table 4. The weights computed in this fashion are representative of the average weight of a catalog position and are not representative of the weight of a single observation. One of the most striking features of the table is the difference in the weight of the right ascension and declination measurements. In many cases, the declinations are determined with less than half the weight of the right ascensions, and with one minor exception, one never sees a declination weight equal to or greater than the corresponding right ascension weight. From this, it is obvious that transit circle measurements in declination, even when performed differentially with respect to an existing catalog system such as the FK4, are not as precise as the right ascension measurements. The differential character of the measurement would tend to remove the influence of effects due to instrumental flexure and anomalous astronomical refraction, leaving the circle, circle detector and diameter correction errors as the dominant factors contributing to errors in the

TABLE 4

WEIGHT OF AN AVERAGE CATALOG POSITION FROM THE FIRST ITERATION

MEAN ERROR OF UNIT WEIGHT = $0^{s}.030$ = $0''.45$

TOP - RIGHT ASCENSION; BOTTOM - DECLINATION

	ABB	BORD	BUCH	CAPE1	CAPE2	LEO7"	NIC	PERTH	SANF	SANJ	SANT	PULK	TOKYO	WAS6"	
+5 0	8.82 4.59	16.00 10.33	12.76 6.25			5.67 3.83		9.18 4.18						7.58 5.61	+5 0
0 -5	7.44 3.00	11.11 7.01	6.69 3.24			5.85 3.24	3.79 1.98	9.18 3.52						7.58 4.59	0 -5
-5 -10	4.80 2.78	10.87 6.25	6.25 3.52			6.15 3.24	3.70 1.98	10.87 4.59						7.17 4.59	-5 -10
-10 -15	5.24 2.78	9.18 5.06				5.49 3.00	3.23 1.75	9.18 4.18	7.17 3.24				5.58 2.78	5.33 3.24	-10 -15
-15 -20						4.80 3.00	2.84 1.65	8.01 3.24	6.05 3.00				4.22 2.58	4.34 2.58	-15 -20
-20 -25						8.82 5.61		9.37 4.18	4.59 2.78				3.19 1.98	3.31 2.11	-20 -25
-25 -30						9.57 5.61		8.65 3.83	3.08 2.25		6.46 2.41		2.84 1.48	2.63 1.40	-25 -30
-30 -35					6.69 4.18	10.87 6.25		11.62 4.59			8.32 2.78				-30 -35
-35 -40					6.69 5.61	12.76 7.91		16.00 6.25			11.89 3.83				-35 -40
-40 -45				5.01 5.06		12.17 9.00		17.36 6.25		4.59 3.00	11.89 3.52				-40 -45
-45 -50				5.33 5.06		11.62 7.91		18.90 5.61		5.41 3.24	11.11 3.24	9.37 3.52			-45 -50
-50 -55				4.34 4.18	5.85 3.24	8.82 7.01		16.89 5.61		4.28 3.00		7.72 3.24			-50 -55
-55 -60					4.87 3.24	11.89 7.01		16.00 4.59		3.90 2.41		8.01 3.24			-55 -60
-60 -65					4.73 3.24	9.97 6.25		12.76 5.61		3.01 2.78		8.82 3.83			-60 -65
-65 -70						11.62 5.61		14.06 3.52		4.73 2.25		11.62 3.83			-65 -70
-70 -75						11.89 5.61		18.37 4.18		6.81 2.25		10.63 3.83			-70 -75
-75 -80						8.82 6.25		14.06 6.25		5.17 2.78		10.63 3.24			-75 -80
-80 -85						8.48 4.59		7.86 4.18		4.34 3.00		7.72 4.59			-80 -85
-85 -90						7.17 4.18		7.30 3.00		2.69 1.65		8.82 3.24			-85 -90

measurement of declinations.

It is particularly interesting to look at the weights of the results from the Perth transit circle because this telescope was the only one not used in a visual mode. A photoelectric micrometer with a multi-slit modulating grid was used at Perth (Høg 1972). As expected, the Perth results show extraordinarily high weight in right ascension when compared with the visual micrometer results of all other participants. On the other hand, the Perth declinations have substantially less weight than one might expect. This was probably caused by a loose bolt in the micrometer which, as described by Høg (1976), was not discovered until a significant fraction of the program had been completed.

7. FURTHER CONSIDERATIONS

At the present time, it is necessary to consider that the question of outlier identification has not been addressed, except for a few (O-C)'s which were so very different from the others corresponding to the same star that they could not be retained in the data basis without doing obvious violence to the computation of the average (O-C). There also remains the necessity of making a comparison between the U.S. Naval Observatory data base and that of the Pulkovo Observatory where a collaborative effort with the Naval Observatory is in progress to jointly produce a compiled SRS catalog. This collaborative effort involves the exchange of data, intercomparison of participant systems and quality of the reduction to the system of FK4, the identification of a mutually agreed upon list of outliers, and a study of the random differences between the (O-C)'s of corresponding observatories. When the collaborators agree that all the comparisons are satisfactory, then it is our expectation that the two versions may be combined with equal weight and the results of that effort presented as the single agreed upon version.

9. CONCLUSIONS

From the foregoing, it is not too early to conclude that when the work of the compilers has been finished, positions in the SRS catalog will have average mean errors of about 0.08 arcsec in right ascension, and 0.10 arcsec in declination.

10. REFERENCES

Corbin, T. 1974 Astron. J. 79, 885-889
Corbin, T. 1978 IAU Coll. No. 48, F.V. Prochaska and R.H. Tucker,Eds. 505
Høg, E. 1972 Astron. & Astrop. 19, 27-40
Høg, E. 1976 Abhandlungen aus der Hamburger Sternwarte, Band IX
Lederle, T. 1978 Centre Donnees Stellaires Inf. Bull. No. 14, 62-68
Scott, F.P. 1967 Progress Report to IAU Commission 8, Positional Astronomy
Smith, C. 1980 Mitteilungen der Astronomischen Gesellschaft Nr. 48
Nicholson, W., Penston, Margaret J., Murray, C.A., and de Vegt, Chr. 1984 Mon. Not. R. Astr. Soc. 208, 911-923 and Microfiche MN 208/2-6

THE EXTENSION OF THE FUNDAMENTAL SYSTEM TO FAINTER MAGNITUDES

Thomas E. Corbin

U. S. Naval Observatory, Washington D.C.

ABSTRACT: During the past 100 years the fundamental system has developed from a small list of 623 bright stars to one of 1535 stars. The compilation of the next fundamental catalog, the FK5, will involve the addition of some 3200 stars and will extend the system to magnitude 9.0. A number of criteria were used for selecting the 2000 stars in the seventh to ninth magnitude portion of the extension, the Faint Fundamentals. These criteria include distribution in magnitude and spectral type, quality of the observational histories, mean errors of the proper motions and distribution over the sky. These criteria have proven difficult to satisfy, especially at the faint end of the magnitude range and in the Southern Hemisphere.

INTRODUCTION

Among Professor Fricke's many accomplishments has been the planning for a new fundamental catalog that will not only be considerably more accurate than the existing one, "The Fourth Fundamental Catalog (FK4)", Fricke and Kopff (1963), but will also greatly improve the density and magnitude range. The improved density will result from the addition of new stars to the fundamental system. The extension of the magnitude range will be due to the inclusion of stars of the seventh, eighth and ninth magnitudes. This expansion of the current fundamental catalog is the latest, and largest, in a series of efforts to make the fundamental system meet the requirements of astrometry as well as astronomy in general. These requirements include higher star densities for more detailed systematic comparisons as well as fainter stars to avoid extrapolations of four and five magnitudes in making the comparisons. Fundamental catalogs have been compiled for a number of specific purposes. However, of these catalogs only one, the "Fundamental Catalog" (FC), Auwers (1879 and 1883), has been followed by others that represented expansions and refinements of the original catalog.

Celestial Mechanics **37** (1985) 285–298.

I. THE FC AND ITS SUCCESSORS

The FC was compiled to provide a systematic basis for reductions only from north of -32 degrees declination. It has a total of 623 stars whose average magnitude is 4.1. The distribution of the FC stars is shown in Figure 1. At the time the FC was compiled

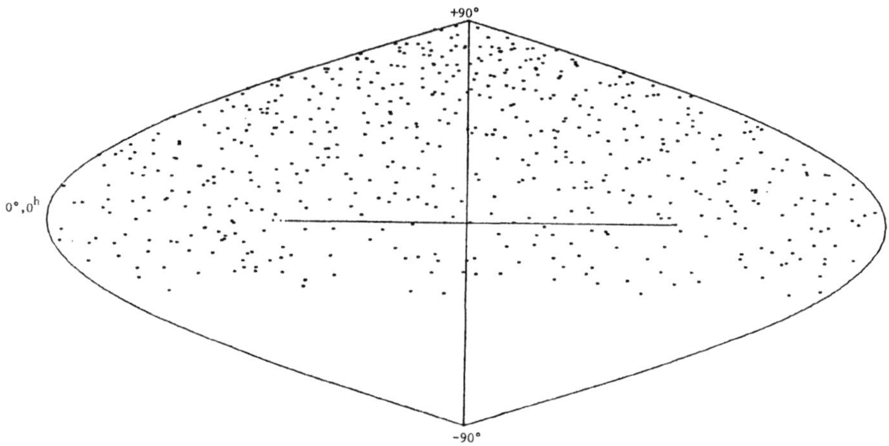

Figure 1. The stars of the FC plotted by position. The center of the plot is ($12^h.0$, $0^\circ.0$).

only the brightest stars had been well observed. Thus it was not possible to determine good positions and proper motions for stars fainter than the sixth magnitude. The resulting distribution of magnitudes is shown in Figure 2.

Both Boss (1903) and Kustner (1902) realized that the observations of the nineteenth century were badly affected by instrument dependent magnitude equations. This was especially true in right ascension. Given this situation it is not really possible to reduce catalogs of eighth and ninth magnitude stars to a system that is mostly composed of stars of the fifth magnitude and brighter. The "Neuer Fundamental Catalog des Berliner Astronomischen Jahrbuchs" (NFK), Peters (1907), improves on the FC system in that much more data was used to compile the positions and motions and corrections for magnitude equation were applied. Also, the system covers the whole sky, as is shown in Figure 3. However, Figure 2 shows that the range of magnitudes is essentially the same as in the FC.

The first version of the "Dritter Katalog des Berliner Astronomischen Jahrbuchs" (FK3), Kopff (1937), has essentially the same stars as the NFK. Again, there had been considerable improvements made in the system

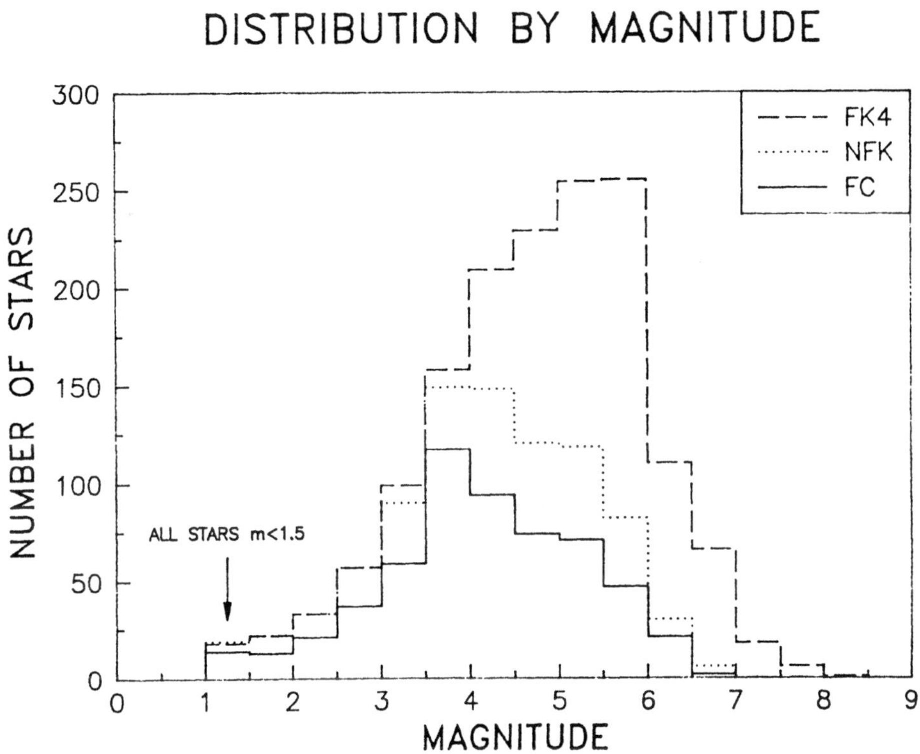

Figure 2. Distribution of the magnitudes of the stars in the FC, NFK and FK4.

of positions and motions, but the problem remained that the fundamental system was much brighter than most of the stars for which it served as a reference frame. This situation was improved somewhat by the additional stars, "Die Zusatzsterne fur die Epoche 1950" (Kopff, 1938) that were added to the FK3. These stars, whose distribution of magnitudes is shown in Figure 2, increased the number of stars mainly in the 4.0 to 7.0 range, but very little fainter than that. The average magnitude of the fundamental system was now 4.8 and had virtually no stars for applications fainter than seventh magnitude. The latest catalog in this series is the FK4, which represents an improvement in the positions and motions of the FK3 stars. However, the FK4 contains no additional stars, and so the FK4 has the same distribution in magnitude as the FK3. The positions of the FK4 stars are shown in Figure 4.

Thus the problem of the fundamental system being considerably brighter than many of the objects that must be referred to it still troubles astrometric work. As Fricke and Gliese pointed out (1968), the next catalog in the series, the FK5, must contain additional stars to extend

II. OTHER FUNDAMENTAL CATALOGS

There have been several other fundamental catalogs compiled in addition to the FK series. Two of these, the "Preliminary General Catalog of 6188 Stars for the Epoch 1900" (PGC), Boss (1910), and the "Catalog of 5,268 Standard Stars, 1950.0 Based on the Normal System N30" (N30), Morgan (1952), have several times the number of stars found in the FK4. The PGC is an extension of the fundamental system in Boss's "Catalogue of 627 Principal Stars." As Figure 5 shows, the PGC does contain a number of stars fainter than magnitude 6.5, but only a small number of the stars are fainter than 7.5 and 89% are brighter than 7.0. Although the PGC was used as a system of reference for a number of years, its primary purpose was for the formation of the "General Catalogue of 33342 Stars for the Epoch 1950" (GC), Boss (1937). In spite of his efforts to make the PGC reach fainter magnitudes, Boss was forced to extrapolate to as faint as magnitude 10.0 in determining magnitude equations of the component catalogs of the GC. At the time PGC was compiled there just were very few stars fainter than magnitude 7.0 with good observational histories. By the time Morgan compiled the N30, the situation had improved somewhat, as can be seen in Figure 5. Some 28% of the N30 (1480) stars are magnitude 7.0 or fainter, and 6% are 8.0 or fainter. Compared to the other fundamental catalogs this was a real improvement in magnitude range. In fact, at one time the N30 was considered to be a possible competitor to the FK series. However, there have been no direct descendants to the N30, and in recent years the FK4 has become the sole astrometric reference frame.

III. SOURCES OF ADDITIONAL FUNDAMENTAL STARS

Since the FK series has clearly become the international standard, any extension of the fundamental system to fainter magnitudes must be made by addition to the stars of the FK4. This approach was addressed by Kopff with the publication of the "Supplement-Katalog des FK3 (FK3 Supp)." The FK3 Supp was compiled mainly from non-FK3 stars in the N30, and, as Figure 5 shows, is composed almost entirely of stars in the magnitude range 4.0 to 7.5. The stars chosen were those that had very good observational histories but that needed additional observations in order to become fundamental stars. These goals were assured of success by the publication of the list reduced to the new fundamental system, FK4. The "Catalogue of 1987 Supplementary Stars to FK4 for the Equinox and Epoch 1950.0 (FK4 Sup)" (Fricke et al, 1963), appearing at the same time as the FK4, identified the list with the future of the fundamental system and thus served as an impetus to observers to give these stars special attention in their programs.

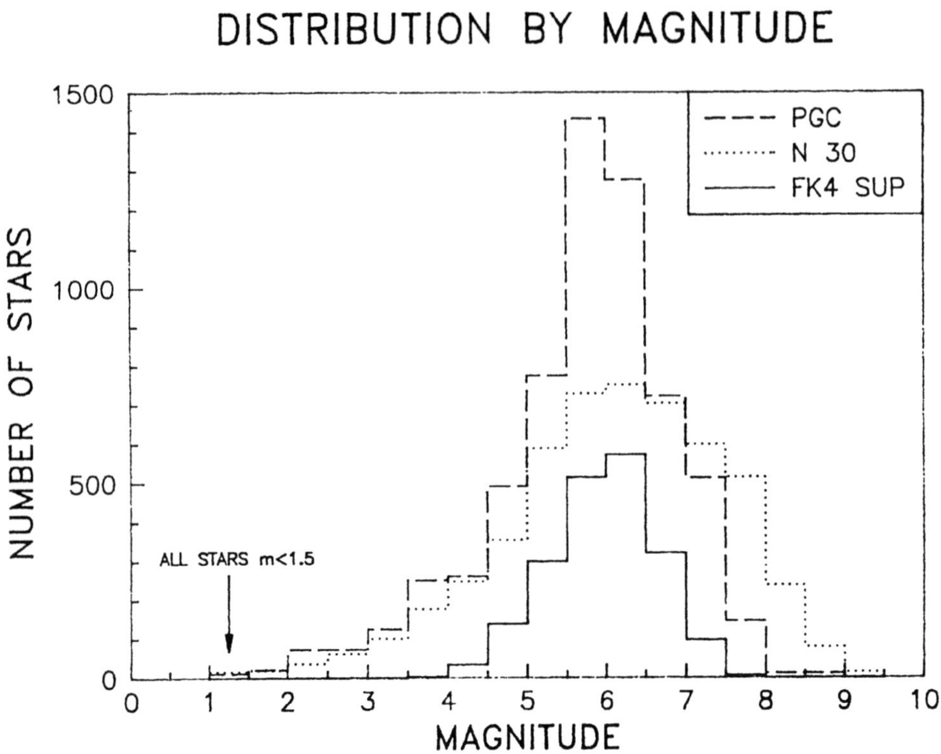

Figure 5. Distribution of the magnitudes of the stars in the FK4 Sup, N30 and PGC.

Figure 5 shows that 55% of the FK4 Sup list is between magnitudes 5.5 and 6.5, and an additional 16% is between 6.5 and 7.0. However, only 5% of the stars are fainter than magnitude 7.0. Thus while the addition of the Sup list to the fundamental system would effectively extend it to magnitude 7.0, there would be little or no coverage fainter than that. Many applications need the fundamental reference frame to be represented at the eighth magnitude or fainter. None of the catalogs in Figure 5 satisfies this requirement. The N30, with about half a percent of its stars 8.0 or fainter, has the best density of the fainter stars. The successor to the PGC, the GC, has 33,342 stars, and many of these are fainter than magnitude 7.0. However, although 22% of the stars are fainter than 8.0, the stars in the GC were selected on the basis of observational history, and this has led to a rather uneven distribution of the fainter stars.

The solution to the problem first took form with quite a different goal in mind. The stars of the AGK3R (Scott, 1963) and the SRS (Scott, 1962) were originally selected with the intention that they would be observed by transit circles in conjunction with astrograph programs. The resulting AGK3R and SRS catalogs would then provide reference stars for the astrographic plates. These two catalogs together were seen to have a wider application by Scott who proposed (1962) that these lists could be combined into a global reference system of almost 40,000 stars that would replace the GC for those applications that required higher densities of stars than found in the fundamental catalogs. The combined AGK3R and SRS stars have come to be known as the "International Reference Stars" (IRS). Since both the AGK3R and SRS were selected in a way that gives very evenly distributed stars (about one every square degree), contains the majority of their stars in the magnitude range 7.0 to 9.0 and gives careful attention to observational history, the IRS list provides an ideal source for additional fundamental stars.

IV. THE FK5 AND THE FAINT FUNDAMENTALS

In 1968 Fricke and Gliese proposed the formation of a new fundamental catalog. They knew that by the time a successor to the FK4 could be prepared the effects of errors in the FK4 proper motions would seriously affect the accuracy of the positions. (Most of the FK4 stars need 60 to 75 years of proper motion applied to bring their positions to the current epoch.) In planning for the FK5 the question was "What stars should be included?" It was obvious that at least some of the FK4 Sup list could be used as these were regularly being given preferential treatment in observing programs. However, as was shown above, the inclusion of the Sup stars could extend the effective magnitude range of the fundamental catalog by only one magnitude. Prof. Fricke saw that something more was needed and in 1973 he proposed that not only Sup stars should be added, but that also a number of IRS stars had to be included. A plan evolved whereby approximately 1200 Sup stars and 2000 IRS stars would be included in the FK5. This bold stroke meant that not only would the FK series receive the largest addition ever to a fundamental catalog, some 3200 stars, but also would be given an extension in magnitude range that would make the new fundamental catalog usable down to magnitude 9.0 without extrapolation.

Such a large addition to the fundamental system as the one that had been proposed requires considerable planning. Naturally the stars that have the best histories of observations are generally the most desirable. However, the needs of modern astrometry require that other factors be taken into consideration. First, a look at Figures 1, 3 and 4 shows that a selection of fundamental stars that has been made mainly on the basis of observational history will not necessarily lead to a good distribution over the sky. The FC and NFK in particular have noticeable variations in the densities of the stars in various parts of the sky. (The low density in the center of the NFK plot is real.) The FK4 is somewhat better. The addition of the Zusatzsterne made the distribution much smoother. The higher densities at the poles were deliberate and are desirable. However, the FK4 has some regions of low star density and others where the stars clump. Second, the distribution of the magnitudes of the extension stars should be compatible with the FK4. That is to say, the FK5 should evenly extend the distribution of the FK4 magnitudes. The third consideration comes from looking at Table 1.

Each of the catalogs in the FK series shows a good representation of spectral types. This is important for a variety of applications such as questions of color terms on astrographic plates and refraction studies. A good distribution of spectral types must be present in the extension stars as well. Fourth, the visual double stars should generally be avoided. These stars can cause problems on both photographic plates and photoelectric detectors. The fifth point is that observational histories of the stars should be examined for catalog positions in absolute catalogs. This is a consideration apart from the general observational histories. A star can appear in a number of differentially observed catalogs and yet have few absolute positions in its history. A good position and proper motion can be determined for such a star and yet it cannot contribute to the formation of a fundamental system. (See Corbin, in press.)

Table 1

DISTRIBUTION OF SPECTRAL TYPES IN THE FK SERIES

Catalog	Spectral Type					
	O&B	A	F	G	K	M
FC	16%	26%	14%	9%	31%	4%
NFK	20	22	15	8	31	4
FK3/FK4	17	23	15	10	31	5
FK4 SUP	13	24	14	11	34	4

V. Selection of the Faint Fundamentals

The selection of 3200 additional stars for the fundamental system is a large undertaking. This is more than twice as many stars as there are in the FK4. Therefore in order to assist Prof. Fricke and his colleagues in this part of the work to prepare the FK5, the U. S. Naval Observatory agreed in 1983 to compile a recommended list of Faint

Fundamentals based on the IRS work that has been done there (Corbin, 1978). The selection of the FK4 Sup stars that will be included in FK5 is being done at the Astronomisches Rechen-Institut. There are some distinct differences between selecting the Sup stars and the Faint Fundamentals. First, the Sups as a group have been well observed, whereas the stars of the IRS have widely varying histories. Some IRS have appeared in two dozen catalogs, while others have only the AGK3R or SRS catalog position. Some 60% of the Sup is to be included in the FK5, while only 5% of the IRS will be. Thus choosing the Sup stars will essentially be a process of eliminating the 40% that will not be part of the FK5. Selecting the Faint Fundamentals means in each case picking one out of many. Finally, the Sup stars are generally well represented in the absolute catalogs whereas a number of IRS stars have no absolute catalog positions at all.

A selection of proposed Faint Fundamentals has been completed. Corbin (1985) has reported on this work for the northern portion of the list, but as the work progressed some additional steps were found to be necessary. The process used is therefore outlined here:

A.) The Faint Fundamentals had to be chosen from the IRS stars with the best observational histories and best determined proper motions. The system of positions and motions on the FK4 system has been completed for the northern IRS, but awaits the completion of the SRS for the southern portion (Hughes and Smith, in press) Therefore a provisional SRS was formed by combining the Washington El Leoncito results (Hughes, Smith and Branham), the W5-50 (Hughes and Scott, 1982) and the Perth 70 (Hog and von der Heide, 1976), each reduced to FK4. This provisional SRS was combined with 97 other southern catalogs reduced to the FK4 system, and resulted in mean positions and proper motions for 15,182 of the southern IRS stars that have three or more catalog positions.

B.) The basic list of candidate stars was compiled from the northern IRS by picking those stars with seven or more catalog positions. In the south, however, this was not possible. North of -20° there is not a problem. The stars have good observational histories, and there is a good list of candidates with seven or more catalog positions. South of -20° the situation deteriorates until in the -50° to -70° zone the candidate list must include stars with only five catalog positions.

C. The sky was divided into blocks of 22 square degrees each. One star was chosen from each block. Initially this gives fewer than the desired 2000 stars (about 1875), but it allowed later flexibility in adjusting the list. The blocks insured that the basic list would be evenly distributed over the sky.

D.) As selection progressed the distributions of the magnitudes and spectral types were checked within each 10° zone of declination. The blocks with the fewest candidates were always done first, and this allowed adjustments of the distributions when the more populous blocks were reached.

E.) The stars selected were matched against the Washington Double Star Catalog (Worley, 1985) and the FK4 Sup. (Some IRS stars are in the FK4 Sup.) In general, double stars for which $m_2 - m_1 < 4.0$ were rejected, although some very close pairs were retained. Most of the selected stars that are also FK4 Sup stars and are brighter than magnitude 7.0 were rejected.

F.) Stars within one degree of an FK4 star were replaced. A few exceptions were made for stars with very good observational histories, although in each case the star had to be 2000" or more from the nearby FK4.

G.) If the mean errors of the proper motion of a star exceed 0ʺ50 in either coordinate, then that star was replaced if possible.

H.) The regions from +80° to +90° and -80° to -90° were given extra stars. This is consistent with the distribution of stars in the FK4, for example, and is needed for such applications as catalog comparisons and providing azimuth stars for transit circles.

I.) Besides the overall observational history, the number of absolute catalogs in which each star appeared was taken into account. This was especially true in the northern hemisphere where absolute observations were not made of some of the stars during the AGK3R program.

J.) Finally, the areas of the spaces between the stars were computed, and an additional star was selected for any area that was 30 square degrees or larger. Plots were then made of the stars and visually inspected for areas of either higher or lower than average density.

The process just described has produced a list of 2072 proposed Faint Fundamentals. The characteristics of this list are shown in Table 2. It had been our goal to select a list that would have approximately equal numbers of stars in the early (O, B & A), middle (G & K) and late (K & M) spectral types in each of the 10° zones mentioned above. Also we intended that about 10% of the list would be in the 6.5 to 7.0 magnitude range, each of the half-magnitude intervals from 7.0 to 9.0 would contain 20% of the list and 10% would be magnitude 9.0 and fainter. Not only were proper motions with mean errors in excess of 0ʺ50/cent in either coordinate avoided whenever possible but also errors in the 0ʺ40/cent to 0ʺ50/cent range were considered undesirable. Table 2 shows that none of these goals was fully realized. First, it must be realized that selecting stars fainter than about magnitude 8.5 is generally not compatible with restricting the list to stars that have good observational histories and good proper motions. This is especially true in the Southern Hemisphere. The southern part of the IRS also presents considerable difficulties in selecting a good distribution of spectral types. South of -30° fully 65% of the IRS stars are either type K or M. Only 5.6% are types O, B

TABLE 2

CHARACTERISTICS OF THE FAINT FUNDAMENTAL LIST OF 2072 STARS

MAGNITUDE		SPECTRAL TYPE	
< 6.5	0.7 %	B & O	6.4 %
6.5 – 7.0	8.5	A	19.6
7.0 – 7.5	21.0	F	18.5
7.5 – 8.0	22.9	G	19.2
8.0 – 8.5	21.7	K	31.6
8.5 – 9.0	18.7	M & OTHER LATE TYPES	4.8
9.0 – 9.5	6.6		

INTERVALS OF MEAN ERROR OF PROPER MOTION ("/CENT)

	0.0/0.1	0.1/0.2	0.2/0.3	0.3/0.4	0.4/0.5	0.5/0.6	>0.6
RA	2 %	18 %	38 %	29 %	13 %	0 %	0 %
DEC	2	14	36	33	16	0	0

AVERAGE MEAN ERRORS OF THE PROPER MOTIONS

RIGHT ASCENSION: 0.29 ARCSEC/CENTURY
DECLINATION: 0.31 ARCSEC/CENTURY

or A. The magnitude distribution is not helpful in this part of the sky either, since 67% of the stars are between magnitudes 7.5 and 8.5.

The list in Table 2 is not final. A comparison of the magnitudes available for the IRS stars with modern photoelectric values has shown that the magnitudes at the faint end of the list in particular are quite inaccurate. We will therefore use the improved system of magnitudes prepared by the CDS, Strasbourg (Egret, 1983), to adjust the faint end of the list. The list will then be forwarded to Prof. Fricke and his colleagues at the Astronomisches Rechen-Institut for final consideration and incorporation into their work to prepare the FK5.

VI. Conclusion

The 2072 proposed Faint Fundamentals when combined with the current stars of the FK4 and about 1200 of the FK4 Sup stars will result in an FK5 whose magnitude distribution will be approximately that shown in Figure 6. The stars are rather evenly distributed between magnitudes

5.0 to 9.0. This is something never before available in a fundamental catalog. The FK5 is thus the logical result of a century of adjustments and extensions to what was initially a restricted and very bright list of stars. The greatest of these additions, however, is the one that Prof. Fricke so clearly saw as necessary for both the current and future needs of astrometry. It will be to him that future users of the fundamental system will owe their thanks for so greatly extending the magnitude range and density of the fundamental stars.

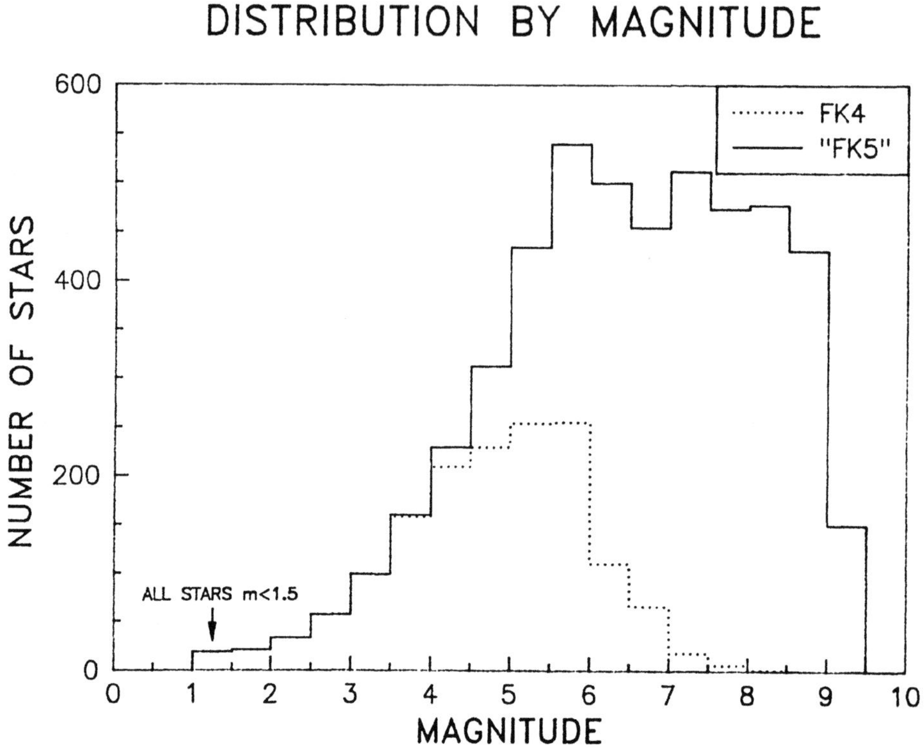

Figure 6. A comparison of the distribution of magnitudes in the FK4 with the predicted distribution in the FK5 (FK4 + 1200 FK4 Sup stars + the proposed Faint Fundamentals).

References

Auwers, A.: 1879, Fundamental-Catalog fur die Zonen-Beobachtungen am nordlichen Himmel, Publ. d. Astron. Gesellschaft 14.

Auwers, A.: 1883, Mittlere Orter von 83 sudlichen Sternen fur 1875,0,

Publ. d. Astron. Gesellschaft 17.

Boss, B.: 1937, General Catalog of 33342 Stars for the Epoch 1950.0, Carnegie Institution of Washington, Publ. No. 486.

Boss, L.: 1903, Astron. J. 23, 83.

Boss, L.: 1910, Preliminary General Catalog of 6188 Stars for the Epoch 1900, Carnegie Institution of Washington, Washington, D. C.

Corbin, T.: 1978, in Modern Astrometry, Prochazka and Tucker (Eds.), IAU Coll. No. 48, University Observatory, Vienna, Austria.

Corbin, T.: 1985, in Astrometric Techniques, ed. H. Eichhorn, IAU Symposium 109, Univ. of Florida, Gainesville.

Corbin, T.: The Determination of Fundamental Proper Motions, in Calibration of Fundamental Stellar Quantities, IAU Symposium 111, ed. D. Hayes, Como, Italy, in press.

Egret, D: 1983, SIMBAD Story, in Bulletin d'Information du Centre de Donnèes Stellaires, No. 24, Strasbourg.

Fricke, W., Kopff, A. et al.: 1963, Fourth Fundamental catalog (FK4), Veroff. Astron. Rechen Inst. Heidelberg, 10.

Fricke, W. et al.: 1963, Preliminary Supplement to the FK4 (FK4 Sup), Veroff. Astron. Rechen Inst. Heidelberg, 11.

Fricke, W. and Gliese, W.: 1968, Highlights of Astronomy, ed. L. Perek, D. Reidel, Dortrecht-Holland, p. 301.

Hughes, J. and Scott, D.: 1982, Results of Observations Made with the Six-Inch Transit Circle 1963-1971, Publ. U.S. Naval Obs. Ser.2, part III.

Hughes, J. and Smith, C: in The Proceedings of the Meeting on the Scientific Aspects of the Hipparcos Input Catalog Preparation, 3 to 7 June 1985, Aussois, France, in press.

Hughes, J., Smith, C. and Branham, R.: Results of Observations Made with the Seven-Inch Transit Circle at El Leoncito, Argentina 1967-1973, Publ. U.S. Naval Obs., in preparation.

Hog, E. and van der Heide, J.: 1976, Perth 70 A Catalog of Positions of 24,900 Stars, Abhand. aus der Hamburger Sternwarte, Bd. IX.

Kopff, A.: 1937, Dritter Fundamentalkatalog des Berliner Astronomischen Jahrbuchs, Veroff. des Astr. Inst. zu Berlin-Dahlem No. 54.

Kopff, A.: 1938, Dritter Fundamentalkatalog des Berliner Astronomischen Jahrbuchs, II. Teil: Die Zusatzsterne, Abh.d. Preuss. Ak. Wiss., math.-naturwiss. Kl. No. 3.

Kustner, F.: 1902, Astron. Nachr., 158, 129.

Morgan, H. R.: 1952, Catalog of 5268 Standard Stars based on the Normal System N30, Astron. Papers Am. Eph. and Naut. Alm., XIII, part 3.

Peters, J.: 1907, Neuer Fundamentalkatalog des Berliner Astronomischen Jahrbuchs, Veroff. d. Kon. Astron. Rechen-Inst. zu Berlin, No. 33.

Scott, F. P.: 1962, Astron. J. 67, 690.

Scott, F. P.: 1963, The System of Fundamental Proper Motions, in Basic Astronomical Data, ed. Strand, Univ. Chicago Press, Chicago, p. 11.

Worley, C. and Douglass, G.: 1985, The new Index Catalog of Visual Double Stars, in Astrometric Techniques, IAU Symposium 109, ed. H. Eichhorn, Univ. of Florida, Gainesville.

PROSPECTS FOR ASTROMETRY WITH THE HUBBLE SPACE TELESCOPE

W. H. Jefferys, G. F. Benedict, P. D. Hemenway and P. J. Shelus
Department of Astronomy
University of Texas at Austin
Austin, Texas 78712
USA

and

R. L. Duncombe
Department of Aerospace Engineering
University of Texas at Austin
Austin, Texas 78712
USA

ABSTRACT. The Hubble Space Telescope (HST), a large optical telescope having an aperture of 2.4 meters and a length of 8.8 meters, is being developed by the National Aeronautics and Space Administration. This telescope will be placed into earth orbit by the space shuttle. Astrometric observations with the HST are made using a Fine Guidance Sensor which is capable of measuring the position of one object relative to another with an accuracy of ±0.002 arcseconds. The astrometric user of HST will be provided with an Astrometric Data Reduction Software package (ADRS). The variety of astrometric problems to be investigated with HST is discussed.

1. ASTROMETRY WITH THE HUBBLE SPACE TELESCOPE.

The Hubble Space Telescope is capable of measuring relative positions of stars within its field of view with an accuracy of ±0.002 arcseconds rms. The magnitude range of HST astrometry is from 9 to 17 without filters, and this may be extended down to approximately fourth magnitude with the aid of filters. Two of the three Fine Guidance Sensors (FGS) are used to maintain the pointing of HST, and the third FGS can be used for making astrometric measurements [1]. One FGS is modified to reduce lateral color effects and is designated as the prime FGS for astrometry.

Three 90 degree sectors of an annulus comprising the outer portion of the HST focal plane are viewed by the three FGS units (Figure 1). Each sector of the annulus (henceforth referred to as a "pickle") is four arcminutes wide and has a maximum chord length of about eighteen arcminutes. The measurement accuracy mentioned above concerns the relative positions of two objects visible in the same pickle.

Figure 2 presents a schematic diagram of an FGS. Each one has an optical beam splitter, two orthogonal Koester's prism interferometers (one for each axis), and associated photomultiplier tubes. The rotation of two beam deflectors, called "star

Celestial Mechanics **37** (1985) 299–305.

selectors", brings light from an object anywhere in the pickle into the five arcsecond square aperture of the FGS detector assembly. When the object is directly on the interferometer axis, the two photomultipliers associated with that axis will give an equal signal and thus the difference of the two signals will be zero. In the case where the object is off axis, the difference produces a non-zero error signal indicating how far off axis the object is. FGS electronics then moves the star selectors to bring the object back onto the interferometer axis, nulling the error signal. The position of one object relative to another in the same pickle can be calculated from the angles θ_A, θ_B and $\Delta\theta_A$, $\Delta\theta_B$, which are given by the star selector positions (Figure 3).

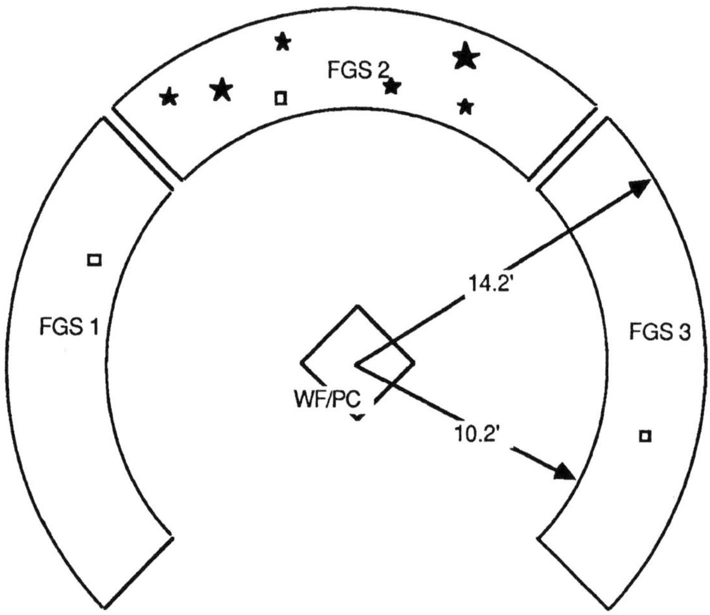

Figure 1: HST focal plane, with locations of FGS fields of view.

The FGS has several modes of operation when used for astrometric measurements. In the "lock-on" mode, the interferometers are nulled successively on different objects within the same pickle. From the readings of the star selectors, the relative positions of the objects in the pickle can be obtained. In the "moving target" mode, which is used for tracking moving objects such as asteroids, the FGS is kept locked on the object and its position is periodically sampled. There is also the "multiple star" mode. In this mode, the target object is allowed to move diagonally across the FGS aperture. The error signal generated in each axis varies as the object moves across the aperture, tracing out the instrumental transfer function (Figure 4).

To reach fainter than 17th magnitude, another mode of operation (the "transit circle" mode) is planned which uses the Wide Field Planetary Camera (WFPC) in conjunction with the guiding FGS units. The WFPC is used to observe the two faint objects in succession. The HST is positioned so that the same two guide stars can be used for guiding during both observations. First, the position of one faint object is measured with respect to the WFPC itself. The HST is then repositioned to measure the other object with respect to the WFPC. The relative positions of the two objects is the difference between the centroided image centers in the WFPC frame plus the difference between the FGS pointings on the same guide stars. This observation type is expected to produce measurements only slightly less accurate than those made in the FGS alone.

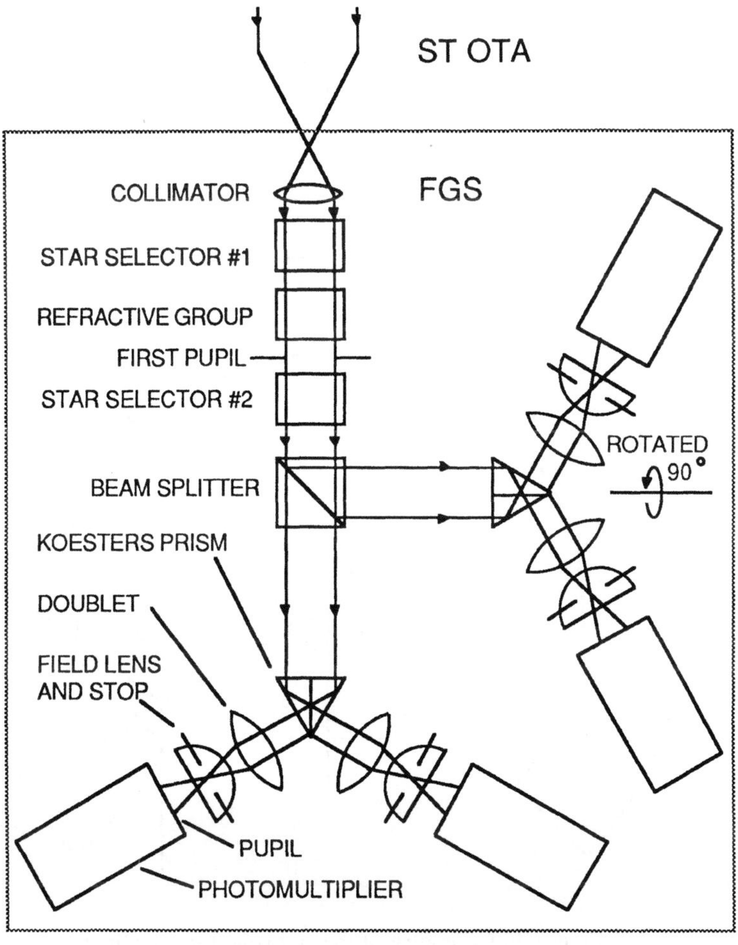

Figure 2: Schematic diagram of the FGS.

It is apparent that there are some restrictions on the use of the FGS for astrometry. Since guidance of the HST uses two of the FGS, only one is left for astrometric measurements. Also, the FGS is not an imaging instrument but is only able to measure the positions of objects within its field of view one at a time. Further, the FGS is not an imaging instrument, but can only sense the presence of an object after moving to its approximate location. Therefore, the area of the pickle must be inventoried in advance, and the schedule of observations must be pre-planned to direct the FGS aperture to each object in turn.

After the launch of HST, the periods of Orbital Verification and Science Verification will allow determination of its capabilities and limitations for astrometry. A calibration field (which has been well charted by ground observation) will be used to determine the zero order plate scale and field distortion effects of the FGS. No star field is available which contains relative positions accurate to ±0.002 arcseconds, over a range of 18 arcminutes. Therefore, the field distortions will be determined by using an overlapping plate technique. Then, using a moving target such as a minor planet, it will be possible to determine the scale factor to within the measuring accuracy of the FGS. Finally, a sample of known single and multiple stars will be observed with the FGS units to verify their transfer functions.

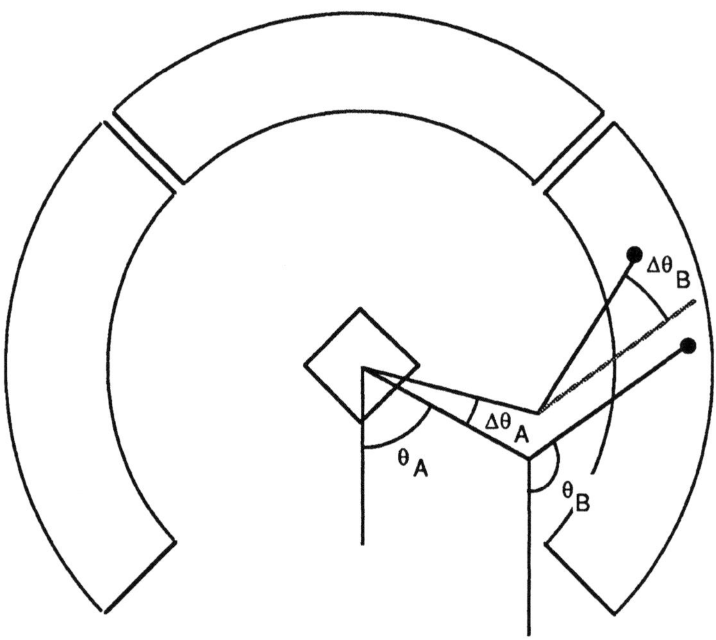

Figure 3: Geometry of the star selector coordinates.

Since the pointing and control of the HST is an engineering function, the data from the FGS will come down the engineering data link. The data from the FGS units will go through the support systems module (SSM), then via the tracking and data relay satellite (TDRSS) to the NASA communication ground station (NASCOM) and on to the payload operations control center (POCC). There the astrometry information is stripped out of the engineering stream and the astrometry information sent on to the Space Telescope Science Institute (STScI) via the science operations ground system (SOGS).

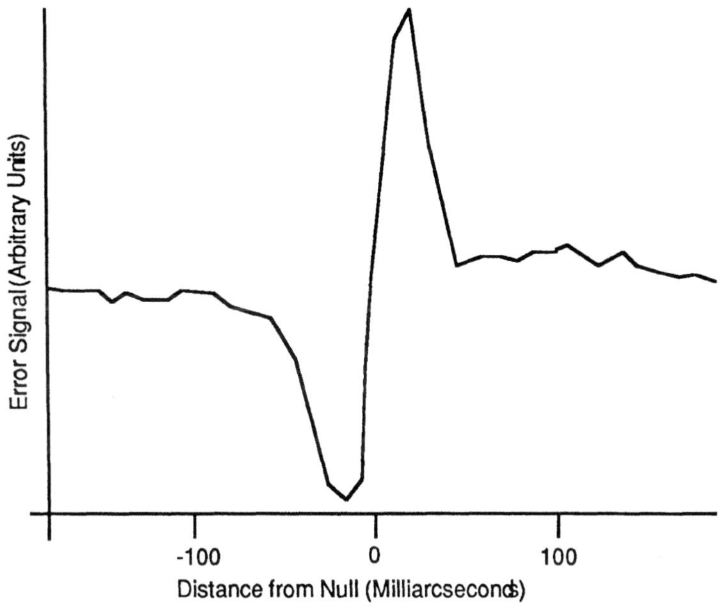

Figure 4: FGS transfer function.

2. REDUCTION SOFTWARE FOR HST DATA

HST astrometric observers will have available to them a library of astrometric data reduction software (ADRS) routines [2]. The user of ADRS will pick from a catalogue those software modules to be applied to the data. Each module performs a predefined operation on the data file and produces in turn a new file on which further operations may be performed. Such standard operations as creation and deletion of files, editing and combining files, copying, saving and renaming files can be performed. Special modules have been created for astrometric data manipulation. The computer operating system allows the user to define and name a sequence of operations in which one module follows another. Assistance to the user is provided by an online "help" program.

Special software modules have been written for astrometric data analysis. Included are modules for conversion among various coordinate systems, for determining the mean position of an object from a sequence of encoder readings, for the correction of the data for various instrumental effects (optical distortion, tilt, etc.) and for physical effects such as differential aberration due to vehicle motion as well as the motion of the Earth. A special module for analysis of transfer function data is capable of providing position angle and separation of double stars which have separations greater then approximately 0.03 arcseconds.

The primary module of ADRS is REDUCE. REDUCE selects a user specified reduction model for the astrometric field being measured. This reduction model may involve any combination of tilt, linear and distortion terms as well as color and magnitude terms. Provision is also made for inclusion of any other terms the user may wish. To specify each model to be used in the reduction, it is only necessary to state the type of observations and the parameters to be used and to specify the equation of condition to be solved. A selection of standard predefined reduction models are also provided for the users convenience. A least squares overlapping plate reduction program is written by REDUCE especially tailored to the specified model [3]. Solution by use of either traditional normal equations or orthogonal transformations of the equations of condition can be specified [4, 5]. Analytical derivatives are calculated for each of the terms in the model using subroutines provided to compute them. The compiler then operates on this program to produce executable code. This REDUCE capability is available with many arbitrary user specified models, including parallax and proper motion solutions, astrometric and visual binary solutions, functional adjustments to single and multidimensional data and orbital calculations. These ADRS programs are now checked out and operating on the VAX computer at STScI, on the University of Texas CYBER and VAX computers, and some modules are being moved to IBM/PC and Apple (Macintosh/Lisa System). These will be available to any user of the HST.

3. HST ASTROMETRIC RESEARCH PROGRAMS

Compared to conventional ground based astrometry, the HST FGS will provide observations which are 2 to 10 times more accurate. This provides the possibility for new and exciting areas of research.

In double star astrometry, the separation range of 0.02 to 1.0 arcseconds is difficult to observe from the ground. Therefore, the statistics of double stars in this range are poorly determined. This situation can be improved, however, if a scan of a star in the third FGS is made routinely every time the telescope is pointed at some primary target for observation by one of the other instruments. Such parallel observations are feasible but their frequency is an operational matter that remains to be determined. In the case of known spectroscopic binaries it may be possible to obtain visual orbits thus providing more knowledge about stellar masses.

It will be possible to measure stellar parallaxes with an accuracy of ±0.001 arcsecond or better, allowing extension of the trigonometric parallax method to such objects as RR Lyrae stars and Cepheid variable stars. It will also provide improvement in the parallaxes of such fundamental distance calibrators as the Hyades. This improved measurement accuracy will make it possible to investigate systematic errors in parallaxes derived from ground observations. It may even be possible to measure the parallaxes of central stars of some planetary nebulae.

HST observations with the WFPC may be used to detect the motions of stars within clusters, and to measure the effects of tidal forces on the cluster and of mass segregation within the cluster. Proper motions of stars in clusters can be detected in a much shorter time than required for ground-based observations, and it may even be possible to measure proper motions in nearby galaxies.

HST observations of solar system objects will provide additional data on the gravitational fields of the planets and satellites. The accuracy of FGS measurement of intersatellite angular distances will provide data good to 5 km at the distance of Jupiter and

10 km at the distance of Saturn. Extended observations of the satellite systems first observed by Voyager may help solve some of the dynamical questions raised by the short duration fly-by data.

Observations with the HST FGS units will make a major contribution to fundamental astronomy in a cooperative program with the ESA astrometric satellite HIPPARCOS. The HIPPARCOS satellite will produce a set of stellar positions, proper motions and parallaxes by effectively observing chords on the celestial sphere. Approximately 100,000 stellar positions and parallaxes will be derived with an accuracy of ±0.001 arcsecond. The reference system for these data will be the HIPPARCOS Instrumental System which will have an unknown "solid body" rotation with respect to a non-rotating reference frame, as well as an unknown zero point. It remains then, to determine the rotation of the HIPPARCOS Instrumental System and to tie it to some fundamental or absolute coordinate frame. The HST because of its capability to measure precise angular distances between objects of disparate magnitude, can be used to tie the HIPPARCOS Instrumental System to: (a) very distant and hence relatively motionless objects such as quasars; (b) an absolute coordinate system derived from radio interferometric observations using radio sources which have discrete optical counterparts, and (c) a dynamical system such as that defined by the motions of selected minor planets [6].

Other possible research programs include the search for planets of nearby stars, the determination of the gravitational deflection of light by Jupiter and the measurement of optical proper motions in quasars.

The Hubble Space Telescope is now scheduled for launch in August 1986. Astronomers are eagerly awaiting this event to utilize its astrometric potential.

4. ACKNOWLEDGEMENTS

We thank our colleagues William van Altena, Laurence Fredrick and Otto Franz for their advice and cooperation in this project. We gratefully acknowledge the support of the National Aeronautics and Space Administration under contract NAS8-32906.

5. REFERENCES

[1] Jefferys, W. H., 'Astrometry with the Space Telescope', *Celest. Mech.* **22**, 175-181 (1980).

[2] Jefferys, W. H., Benedict, G. F., Hemenway, P. D., and Shelus, P. J., 'Space Telescope Astrometry Software', *Bull. American Astron. Soc.* **14**, 581 (1982).

[3] Jefferys, W. H., 'On the Overlapping Plate Method in Astrometry', *Astron. J.* **84**, 1775-1777 (1979).

[4] Jefferys, W. H., 'On the Method of Least Squares. II', *Astron. J.* **86**, 149-155 (1981).

[5] Jefferys, W. H., 'Overlapping Plate Reduction Using Orthogonal Transformations', *Bull. American Astron. Soc.* **13**, 569 (1981).

[6] Duncombe, R. L., Hemenway, P. D., Whipple, A. L., 'Minor Planet Observations and the Fundamental Reference System', *Celest. Mech.* **34**, 19-36 (1984).

RECENT INVESTIGATIONS ON DISK GALAXIES IN MASSIVE HALOS

R. H. Miller
University of Chicago
Department of Astronomy and Astrophysics
5640 Ellis Ave.
Chicago 60637 U. S. A.

ABSTRACT. Numerical experiments undertaken to investigate the longevity and behavior of dark-lane elliptical galaxies are described. This is dynamically the same problem as a disk galaxy in a massive halo. Spiral galaxies are disks from a dynamical point of view. A disk of particles embedded in a self-consistent galaxy provides the basic model used for the experiments. This model is applicable to ordinary disk galaxies if the disk is interpreted as the visible galaxy and the "galaxy" is interpreted as the massive halo thought to be present around disk galaxies. Fully three-dimensional fully self-consistent n-body computer programs that can handle 100,000 particles are used for the experiments. The background galaxy is oblate, and the disk is inclined to the axis of the oblate galaxy, so the disk precesses differentially to produce a warp. A surprising result is that the galaxy center shifted, leaving the disk center orbiting around the galaxy center. This produces interesting phenomena reminiscent of observations in the region of the Galactic center.

1. INTRODUCTION

The dynamics of galaxies is a beautiful subject. Galaxies are beautiful both in photographs and in VLA maps, and they are dynamical objects. Our attempts to study the the dynamical problem experimentally produce strikingly beautiful objects as well as charming dynamical insights. Galaxy dynamics is challenging because one's guesses are so often wrong. This happens, of course, because self-consistent systems have so many options open to them that they usually select one you hadn't thought about. It is exciting as you begin to see which option the system selected and as you try to figure out why it chose that option. Professor Fricke once worked in the dynamics of galaxies, and his interest in this subject is my excuse for discussing a problem a little off the beaten path of the remaining papers of this collection. My appearance on this program bears testimony to the breadth (as well as the well-known depth) of Professor Fricke's interests. It is a privelege to be invited to make this presentation, and I thank both Professor Fricke and the organizers for the opportunity.

This investigation started out as a study of the dynamics of dark lanes in the so-called dark-lane elliptical galaxies. Some totally unexpected and surprising phenomena turned up--phenomena that remind one

Celestial Mechanics **37** (1985) 307–324.

strongly of the very strange things observed at the center of our Galaxy. Our current research is pretty much centered on this investigation, which is far from complete. We are breaking new ground, and we must develop methods to analyze the results and design new experiments to test our hypotheses as we go along. However, the results are so startling and so beautiful that we could not resist discussing them at this festival.

Dr. Bruce F. Smith of NASA-Ames Research Center in California and Dr. Althea Wilkinson of the University of Manchester in England are the team that makes the project go. Dr. Wilkinson originally suggested the problem. Computations for the *n*-body numerical experiments have been carried out at the NASA-Ames.

The prototypical dark-lane elliptical galaxy is NGC 5128, a southern galaxy that is identified with the strong radio source Centaurus A. It is a very strange looking object at first glance, but on closer study it looks like an elliptical galaxy, flattened a little (E1 or E2), but crossed by a bold dark lane along the short axis. The dark lane cuts the visible galaxy image in two. Current theories of the dark lane favor models in which a disk galaxy, containing a lot of dark matter, recently merged with the elliptical (Tubbs 1980, van Albada, Kotanyi, and Schwarzschild 1982). Wilkinson *et al.* (1985) recently tied radio and optical studies of the system together. Several of these dark-lane systems are known. Many are radio galaxies (Kotanyi and Ekers 1979). No others are as near, as bright, or as prominent as Cen A.

Disk galaxies in massive halos present the same dynamical problem, especially if the disks represent but a small part of the total mass. Our Galaxy is thought to be a disk galaxy, and it is thought to have a massive halo. Things we see going on at the center of the models described here might very well occur in the center of the Galaxy. Both "flat rotation curves" (Rubin *et al*, 1985) and warps seen in some disks (Sparke 1984) are interpreted as indicating that a lot of dark matter is out there. The dark matter is dynamically important (after all, its presence is inferred on dynamical grounds), but the stuff of which it is made and its precise form are not known. The simplest assumptions are that it is nearly spherical and that it is made of some kind of particles (stars, black holes, massive neutrinos...?) that are simply not luminous. Dynamically, it acts like an elliptical galaxy. The disks that make up the beautiful spirals and other similar galaxies are simply luminous matter trapped in the potential well provided by the dark matter. The dark matter is usually called a "halo"--a misnomer since it does not glow.

2. THE EXPERIMENTS

Numerical experiments are initial value problems. They are defined by the initial conditions. The same rules are used for carrying out all experiments, and those rules contain as much of the essential physics of galaxies as possible. Once an experiment has started, it runs to "completion" with no interference.

The experiments discussed in this note were started with a disk embedded in an oblate spheroidal galaxy. Both the disk and the oblate spheroid were formed of particles, 100,000 in all. The experiments consist of following the dynamical development of this system for some time, usually for several dynamical times (equivalent to orbital periods in celestial mechanics), by means of a fully self-consistent fully three-dimensional *n*-body integration. The bulk of the mass is in the oblate spheroid, and it represents the elliptical galaxy. The disk represents the dark lane. It contains little mass (1000 particles), and notwithstanding its dominant optical appearance it has little, if any, dynamical effect on the main galaxy.

We don't ask how the disk and the galaxy came to be together. There is little point in worrying about how the system might have been formed if it doesn't live very long or if the dynamics is not compatible with observations of dark-lane ellipticals. The dynamics and longevity are the object of the present investigation.

The "elliptical galaxy" was made by squashing one of our usual n=3 polytropic spherical models so that equidensity surfaces are nested concentric oblate spheroids. The equidensity surfaces were nearly spherical near the center with increasing flattening at greater distances until the outermost parts were flattened nearly 2:1. "Random" velocities within the galaxy were also flattened to help maintain the oblate form. In addition, the galaxy was made to rotate, again to help it maintain its flattened form.

The disk was flat at the start. Its center coincided with the center of the galaxy. Disk particles orbit around the galaxy center so the disk is centrifugally supported against the gravitational forces of the galaxy. The disk rotates differentially. Its surface density falls off away from the center as $1/r$. The disk was initially "cold:" disk particles had no random motions. The analogue in our Galaxy would be that all disk stars were on circular orbits, without epicyclic motion.

The initial orientation of the disk defines the experiment. All the other properties were the same from run to run, so there is only one parameter: the angle between the disk normal and the spheroid's symmetry axis. An angle of zero means that the disk lies in the spheroid's equatorial plane, while an angle of 90 degrees makes the disk lie in a meridional plane, like a "polar ring" galaxy. Experiments were run at 30, 60, and 90 degrees. Sixty degrees is the best guess for the inclination of the dark lane in Cen A (Wilkinson *et al.* 1985).

Disk and galaxy particles have the same mass. They are treated identically in the integration. They are only distinguished by a "flag" that enables the computer programs to tell which is which in the "plot files" that we make to record the progress of an experiment.

The plot files contain the x, y, and z coordinates of 2048 particles, recorded at each integration step. All disk particles were in the plot file (1024 of them); the remaining 1024 particles in the plot file were galaxy particles. Only 1/100 of the galaxy particles are plotted in motion pictures.

Each disk particle precesses with a motion like that of a satellite orbiting an oblate planet. Here we have many satellites, initially forming a flat sheet. Satellites at different distances from the center precess at different rates, so the originally flat disk is soon warped into a curved sheet. Indeed, one finds beautiful warps in these experiments. The disk does not settle into the equatorial plane like Saturn's rings. This system is non-dissipative, and the time scale for the disk to settle is too long.

Once a particle load has been produced, our usual *n*-body integration routines (Miller 1978) take over. Forces are computed from potentials that are tabulated on a cubic lattice within a larger cubic region. There are 64 active grid points along each coordinate direction. Free boundary conditions apply: potentials go to zero at infinite distance. The equations of motion for each particle are integrated by a time-centered leapfrog method. We use one integration step as the time unit for the calculation. The potential solution and the integration are crude compared to those used in celestial mechanics because self-consistency is emphasized. Potentials are much more complicated than those that arise in celestial mechanics, and those potentials must be recomputed at each integration step to keep pace with dynamical developments within the system.

The experiments run are as follows.

Run Name	Inclination of Disk Normal	Run Duration
AWOBL2	30	256
AWOBL3	60	256
AWOBL4	90 (polar disk)	256
AWOBL5	30	768

AWOBL5 is a long run like AWOBL2 except that a different sequence of pseudo-random numbers was used to generate the particle load. An initial crossing time for all runs is 35 steps, so the 256 step runs are a bit more than 7 initial crossing times and the 768 step run is nearly 22 initial crossing times.

3. EXPERIMENTAL RESULTS

For such a simple arrangement, the experimental results are unbelievably rich and complex. A motion picture was shown to illustrate the dynamical responses of the system. This is a powerful way to study dynamical developments. The continuity of development conveys information that it is impossible to communicate effectvely otherwise. Unfortunately, we have not been able to find any way to demonstrate those results on the printed page. The best we can do is to attempt verbal descriptions. Later, we give some details on subsequent numerical studies of the experimental results.

3.1. Shapes

The first section of the film shows scenes of the system at fixed time, rotating the complete configuration to show the three-dimensional form of the galaxy and of the warped disk. The first scene shows the system at the end of AWOBL2, then the second shows the system at the end of AWOBL3, and yet a third scene shows the configuration at the end of AWOBL4. This is followed by three more scenes, each showing a stage in the development of the longer AWOBL5 run. Galaxy particles are plotted in blue and disk particles in yellow. The moving pictures look a bit like a field of spring flowers in Texas, but we rotate the view so you can appreciate the shape of the hill on which the flowers grow. It produces beautiful pictures that can be enjoyed for the fascinating patterns and shapes, quite apart from the dynamics.

General features are pretty much the same in all the experiments, even though details of the disk shapes were different in different experiments. The disk remained sheetlike each time; it did not puff up into a spheroid. It did not develop folds, loops, or short-wavelength oscillations. Nonetheless, the shapes are difficult to describe either in words or mathematically. A warp is noticeable in the central regions first, and then it seems to move outward. At early stages the center of the disk is warped and the outside looks flat. The scenes at the end of experiments 2, 3, and 4, and the scene at step 256 of the longer AWOBL5 run all look like this. Later, the center looks pretty flat and the outside has a nice "hat-brim" warp--sharply upward on one side and downward on the other (relative to the flat central region). Numerical studies of the warped disk show steady precession at all radii, but precession is slower at the outside. The visual impression depends on noticeable distortions, and these lead to a somewhat different description.

Some rings appear in the disks. These are patterns that result from epicyclic oscillations of the very cold disk. They are a set of density waves. Oscillations are set up because disk particles are not initially on strict equilibrium orbits, and the rings appear where lots of particles have piled up because the oscillation phases bring them together. In the temporal films, the rings look a bit like waves in a puddle that propagate outward after a pebble has been dropped into the center. Similar rings appear in other experiments that were run with the self-consistent effects blocked, so self-consistency does not seem to be required to form the rings or to hold them together.

The disk in the "polar ring" experiment AWOBL4 remains nearly planar. There is no warping. The center becomes distorted, however. We return to that later. The center shows bars, rings, and lenses, all in the same system, when viewed from different directions.

3.2. Temporal developments

A second section of the film shows the temporal development of the longer AWOBL5 run from each of three orthogonal view directions. The view direction remains fixed in this section and we watch the system develop dynamically as time progresses. This section is also in color, and it is color-coded the same way as the first: blue for galaxy particles and yellow for disk particles.

The first view is along the spheroid rotation axis. Once the temporal development has been completed, we go back to the beginning of the experiment to show the complete temporal development of the system from a different view direction. All this is repeated yet a third time from the third orthogonal view direction. The rotation axis is vertical on the screen in these last two scenes.

Nothing very startling happens. Of course, we didn't want much to happen. The initial galaxy was not quite in a steady state dynamical equilibrium. It rearranged itself a bit before reaching a steady state. The disk evolves smoothly from its original flat shape to a warped shape.

The center of the disk oscillates within the galaxy. It moves from side to side and up and down. That motion is one of the exciting features, and we next turn to a more careful look at it. A little spiral at the center opens and closes as it orbits. Further, one sees the ring patterns appear and then move outward slowly.

3.3. Oscillations of the Disk Center

The third section of the film was made to study oscillations of the disk center in the AWOBL5 run. The disk center oscillates in other experiments as well. We show them for AWOBL5 because it is a longer run. All disk particles inside a small cubic volume at the center of the configuration are plotted. They are plotted to a scale about 15 times larger than that of the first two sections of the film, so we see a greatly enlarged view of the central portions of the disk. This section of the film also runs 10 times faster than the first section. Things happen very fast in this section. Disk particles are plotted in white against a black background. This section is not in color. Galaxy particles are not shown.

Temporal developments are shown from each of three orthogonal view directions, as they were in the previous section of the film. The first scene shows a view along the spheroid's symmetry axis. Once that first scene is finished, we go back to the beginning to show the entire temporal sequence from an orthogonal view direction, with the same enlarged view and with the same very fast time scale. And we repeat for the third orthogonal view. The disk is instantaneously seen edge-on at the start of this third scene.

Disk particles were loaded into a spiral pattern at the start. That spiral pattern persists at the center, although it becomes quite distorted. The pattern rotates, but it also pulsates. These pulsations are epicyclic oscillations down near the center of the galaxy. A general threshing about soon destroys any reminder of a spiral except for a little patch near the center, but particles remain in stringlike features throughout. Two-body relaxation processes must have been quite effectively suppressed. Otherwise the stringlike features would quickly fuzz up and no strings could be recognized.

The little spiral at the center soon starts to whip about. We made this section of the film so we could watch the center as it whipped about, to watch its motions and the way it changes shape. It whips very fast, but the innermost turn of the spiral retains a recognizable form. That innermost turn of the spiral is a remnant of the initial load, not something that developed through some kind of dynamical process.

Things happen so fast in this section of film that the motions tend to be bewildering on first viewing. However, the center of the disk can be seen to be in orbit about the center of the galaxy. The disk center rotates, precesses, and nutates as it orbits. It is a fantastically complex motion. Nonetheless, the little spiral at the disk center remains recognizably a spiral through all this. The center also whipped about in the polar disk experiment, AWOBL4. "Whipping" is likely to be the process that created the out-of-plane distortions that we mentioned in connection with the first section of this film.

3.4. Numerical Studies

The "whipping about" of the disk center was totally unexpected, at least to us. We were concerned lest it be warning us of some numerical difficulties in the integration, such as a numerical instability. We therefore launched into a fair number of checks to try to determine whether this phenomenon was something numerical or was telling us something about the physics of a disk embedded in a galaxy. As you can imagine, this is not easy. Further, it tends to be done in a state of panic because anything wrong is likely to be something that has been lurking in all your previous work. We concluded that the phenomenon was not of numerical origin and that it was indeed telling us about the physics of disks in galaxies. It would take us far afield to report all the checks and tests. Some interesting physics turned up in these checks. We describe that physics here.

Imagine a small patch at the center of the disk. We define the center by the 64 disk particles that were closest to the center at the beginning of the experiment. The third section of the motion picture showed that these 64 remain together and move together. We can find a best-fit plane through them (in a least-squares sense), and we can discuss the motions of a unit vector normal to that plane. The end of the unit vector moves about on the unit sphere. Those 64 particles initially occupy the innermost 1/16 of the disk's radius, so the patch is pretty small. The sum of the squared residuals from the least-squares process is telling us how fat a blob the 64 particles make. Various noise processes cause individual particles to diffuse about and to fatten, but the patch remains quite thin.

We show the trajectory (throughout the AWOBL5 experiment) of the end of the unit vector in Figures 1 and 2. These are two presentations of the same data. Figure 1 shows the trajectory as seen looking down onto the pole of the unit sphere. The pole is marked by the small cross at the center and the circle represents the equator. The trajectory is the curve that loops and wiggles across the figure, looking like something drawn by a small child. It starts at the right-hand side where the wiggles are small. It starts moving downward, develops some rapidly growing loops, and drifts toward the left. It finally ends with some huge loops at the left. Figure 2 shows the same trajectory in a latitude-longitude plot (equivalent to a Mercator projection). It starts at the left-hand side of Figure 2, where the wiggles are small, and proceeds toward the right. The start at 60 degrees is not an error--the normal to a 30-degree disk strikes the sphere at 60 degrees.

The loops in these two figures represent a free nutation, like that of the spinning earth. They are a physical process. The drift of the loop centers is the precession. Note that the loop centers drift at nearly constant latitude, a signature of precession. Compare the drawings of combined precession and nutation in any mechanics book. Also note that the orbital motion of the disk center does not affect these plots because they refer only to the direction of the disk normal. The complex motions of the disk center seem remarkably well-behaved once we recognize the motion as a combination of orbital motions with precession and free nutation. We return to the orbital, precession, and nutation rates later.

Next, we looked at the responses of the entire disk. Now imagine the disk divided into 16 annular rings, with 64 particles in each ring. Again each ring was defined by a set of particles and these particles were selected on the basis of their locations at the beginning of the experiment. A "bin" or a "ring" is actually a crude term that denotes that particular set of disk particles. This kind of binning is necessary when the entire disk orbits around the galaxy center. It is necessary to allow portions at various radii to orbit at different orbital phases. It makes sense if the particles in various bins do not mix very much. They don't. A plane can be fitted to the particles in each of these rings, again by a least squares process. And again, the squared residuals from the least squares process tell us something about the amount by which the sheet has fattened up or curved. The planar fit is pretty good even to the end of the longest run (AWOBL5) and to the outermost ring. The patch at the center, discussed a few paragraphs back, is the innermost bin.

The set of sixteen latitude-longitude plots for the AWOBL2 experiment is shown in Figure 3. This experiment had a 30-degree disk

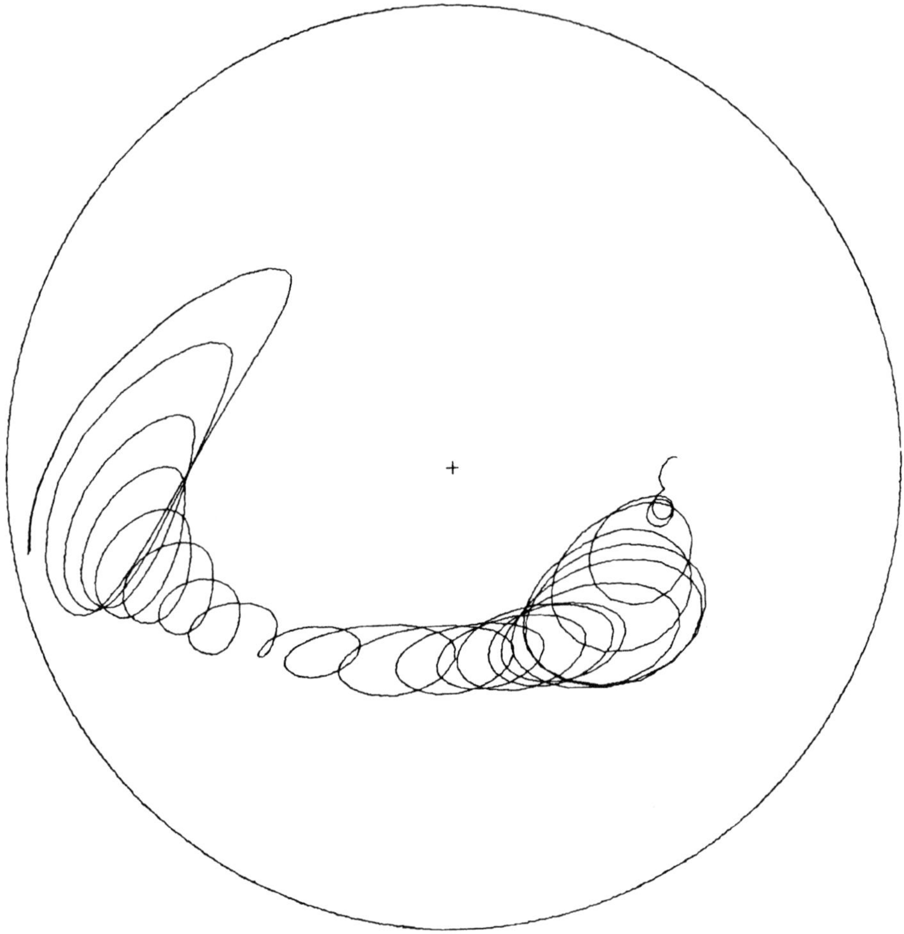

Fig. 1. Trajectory of the normal to a little patch at the center of the disk in experiment AWOBL5, as seen projected onto the equatorial plane of the unit sphere. The trajectory starts at the right and drifts down and to the left. The loops are evidence of free nutation, while the drift represents precession.

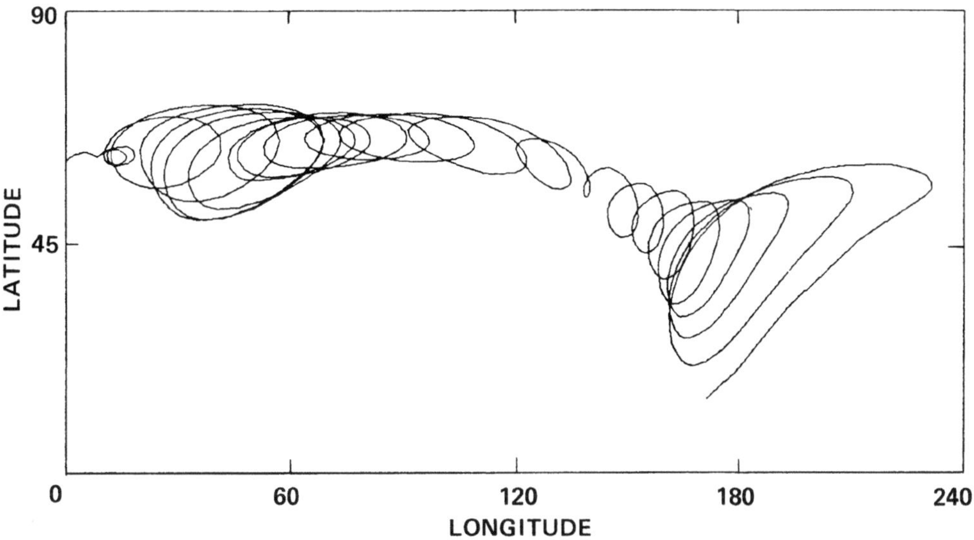

Fig. 2. The same trajectory as shown in Figure 1, this time on a "Mercator projection," with latitude vertical and longitude horizontal. The plot shows the normal to the disk center as it drifts about the unit sphere. The trajectory starts at longitude zero, and it drifts to the right as the experiment proceeds.

Captions for the next four figures:

Fig. 3. "Mercator projection" plots of trajectories of the normals to various annular rings in the disk throughout experiment AWOBL2. Tracks have been vertically displaced for clarity. The trajectory referring to the outermost bin is at the top, and that for the center patch is at the bottom. These show a coherence of nutation from one radius to another.

Fig. 4. Plots of longitude *vs.* time for the 16 annular-ring bins of experiment AWOBL2. Again, the track for the outermost bin is at the top, that of the center patch is at the bottom. Tracks have been displaced vertically for clarity. Nutational motion gives the appearance of outward-propagating waves.

Fig. 5. Same as Figure 3, with the data of experiment AWOBL3. This experiment had a 60 degree disk inclination in contrast to the 30 degree inclination of AWOBL2.

Fig. 6. Same as Figure 4, with data from experiment AWOBL3.

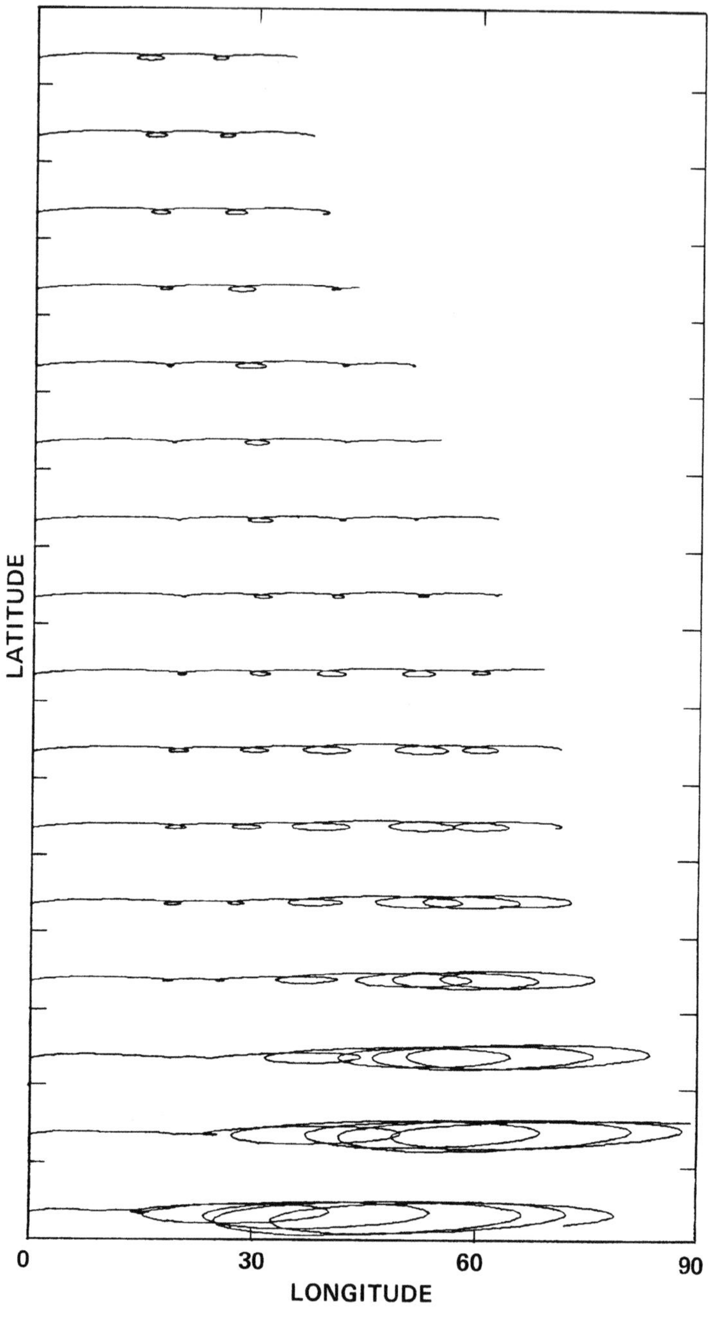

Figure 3

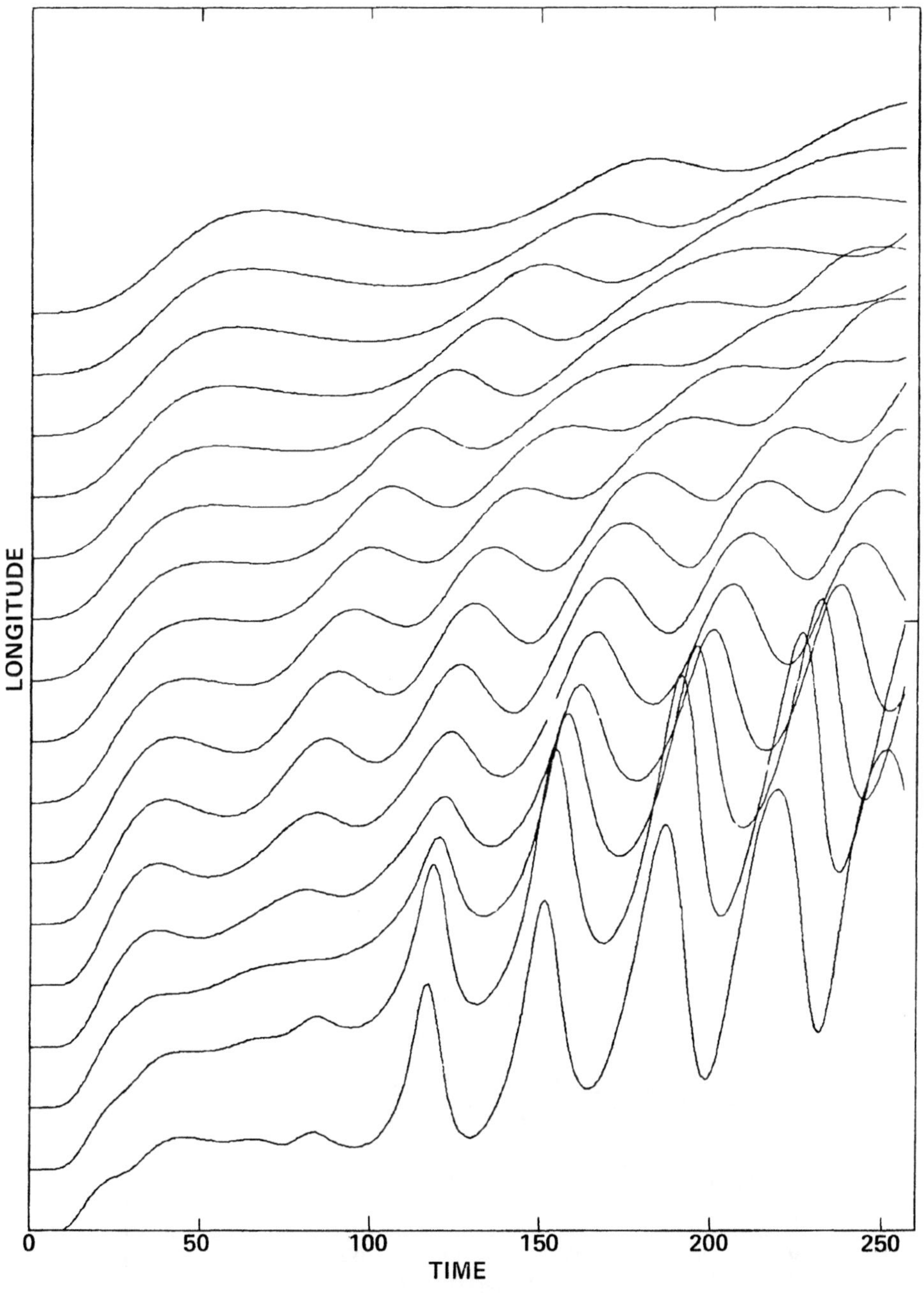

Figure 4

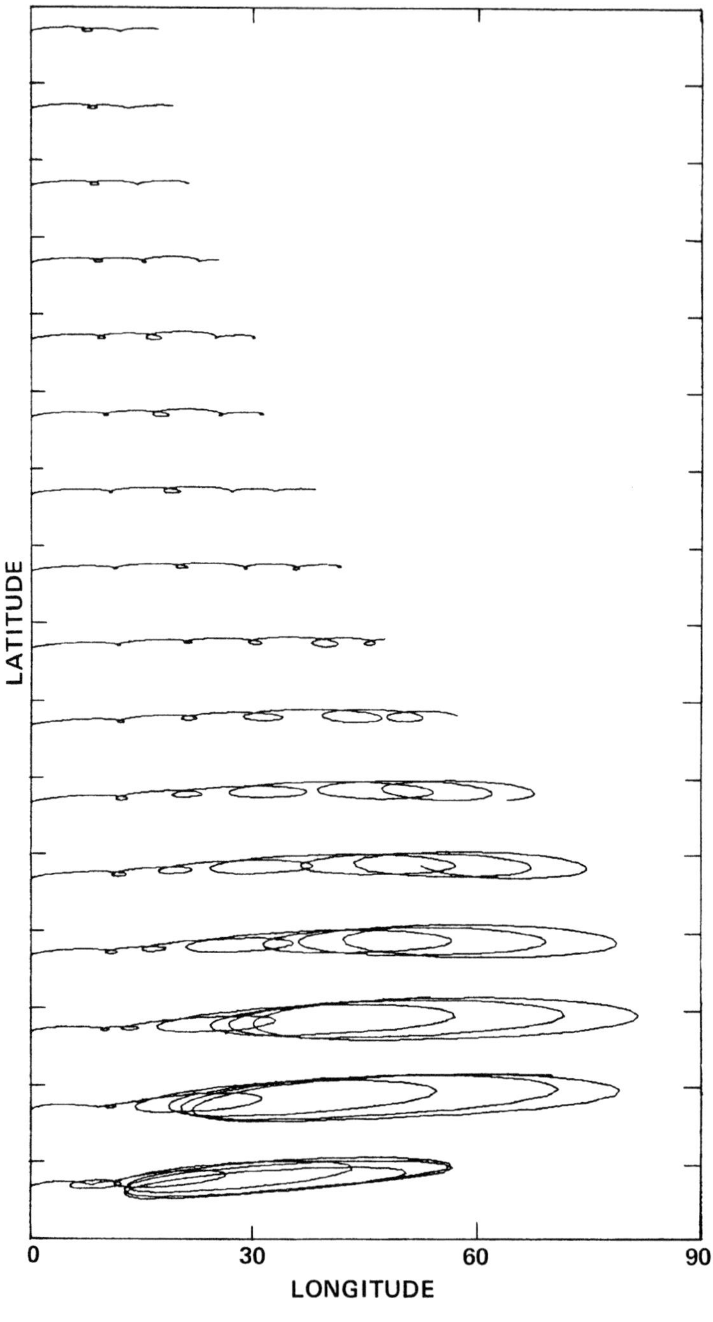

Figure 5

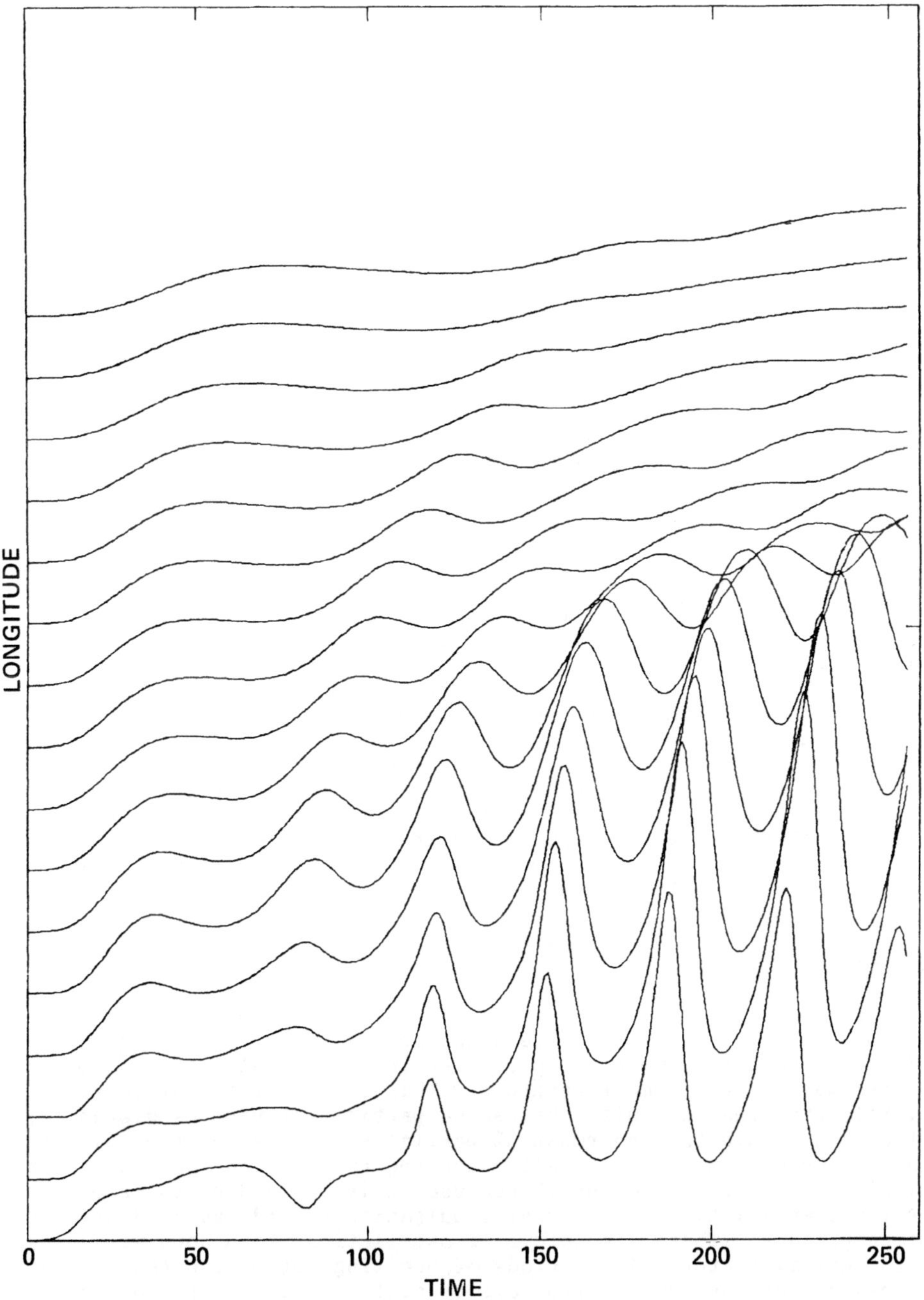

Figure 6

inclination like AWOBL5. The plots are oriented in the same direction as figure 2, but the scales are different. The trajectories have been compressed vertically to get them into a plot and they have been displaced vertically to disentangle them. The trajectory representing particles in the outermost bin appears at the top of the plot and that of the center patch is at the bottom. The trajectories are not as long or convoluted as that of Figures 1 and 2 because this is a shorter run than AWOBL5.

Several features are noteworthy. (1) The total precession angle is smaller for the outer bins than for the inner bins. The total precession angle is about the same for the inner 8 bins. Beyond those inner 8 bins, it decreases outward. The innermost bin (that center patch) precesses less than the second and third bins. (2) No drift in latitude angle is noticeable in the precessional motion. This statement is a little crude because of the compressed scale in Figure 2 and because of the confusing effect of nutation. (3) The loops that represent nutation are present all the way out to the edge of the disk, but those in the outer bins are much smaller. (4) Trends can be seen in the longitudes of these loops in neighboring bins.

The pattern is intriguing. However, the times corresponding to these features in Figure 3 cannot be deduced from that figure. Accordingly we replotted the data in Figure 4 to show the longitude angle as a function of time. The amplitude of longitude swings (in the nutational loops) are large enough to cause some of these tracks to cross, but that does not detract from the message of Figure 4: the nutational patterns appear to propagate outward, with decreasing phase velocity in the outermost bins. Again the track belonging to the outermost bin is at the top and that of the innermost bin is at the bottom, and the tracks have been displaced vertically for clarity.

The same kinds of plots are given for AWOBL3 in Figures 5 and 6. These show that an experiment started with a different disk inclination (60 degrees instead of 30) produced the same kinds of responses. There is nothing special about the 30-degree experiments. The same kinds of plots for AWOBL5 have the same character as well.

A list of the periods of motions we have identified will give some idea of the complexity of the motions.

1.	Radial pulsation	40 time steps
2.	Rotation of the spiral pattern	65-67
3.	Orbiting motion	68 in z, 74 in x and y
4.	Nutation	32, and
5.	Precession	1360.

The rotation of the spiral pattern was determined by following the difference-vector between two particles in the spiral pattern. One of the two was the centermost particle. This figure is the average of several determinations with the second particle at various distances, but always within the innermost 32 particles. It is not the same as a "pattern speed" in spiral density-wave theory. In that parlance, our spiral is a "material feature," because it is defined by the same particles at all times. The radial pulsation period was read off the motion picture. These are epicyclic oscillations seen down very close to the galaxy center. The periods of orbiting motions, nutation, and precession were determined from direct plots. The z-direction for orbiting motion is along the spheroid's rotation axis, while the x- and y-directions lie in the equatorial plane. The period of rotation of the spiral pattern (2) is measurably different from that of the x,y,z

orbiting motions (3). They are not the same. Our accuracy is good enough that we can see a clear difference. The nutation rate is about double the ($G\rho$) time down in the galaxy center, and the orbiting motion is nearly at that rate.

4. PUZZLES

Our attempts to interpret these results raise as many questions as they answer. Many of these newer questions would best be answered by means of numerical experiments, but we have not yet figured out how to design experiments that will yield clear-cut answers for most of them. We can formulate some of those questions in this section.

Most of the questions follow from the rapid motions of the disk center as it orbits around the galaxy center. Our present thoughts about what caused the center to start orbiting are as follows. The initial galaxy was not quite in a steady-state equilibrium, and it reshuffled itself shortly after the beginning of the experiment. Plots of the total kinetic energy *vs.* time verify that there was some reshuffling. It looks as if the center of the galaxy's potential (the place with the most negative values of a smoothed potential) shifted away from the centroid of the total mass and shifted away from the place where we had expected it to be--at the disk center. There is nothing that requires a galaxy's center, as defined by the smooth potential, to coincide with the centroid of the total mass. The magnitude of the shift is not large--it is a fraction of a core radius of the galaxy. It would be less than 100 pc scaled to our Galaxy.

Once the disk found itself in this situation, its center started to orbit around the new potential center. But orbiting induces even more complex motions: the orbit precesses in the oblate potential (with a period given by the difference of the z-frequency from the x- and y-frequencies), while the disk precesses and wobbles as it spins. This is like the wobbling of a spinning plate: the disk normal traces a series of loops on the unit sphere as a result of this wobbling. The wobbling is like free nutation in the earth's motion, so we call it "nutation." The wobbling motion goes on as the disk orbits. The aggregate motion is like the simultaneous orbital motion, orbital precession, body rotation, body precession, and nutation of the earth or of the moon. You all know what a mess that is to describe. All this sounds quite plausible as an explanation in terms of simple mechanics, like the motions of a rigid body.

It brings up the first question, however. The disk center is a group of particles that presumably have fairly weak--even negligible--interaction. Yet they orbit and wobble as if they were almost a rigid body--as if they were a "thing." It is surprising that the disk center should hang together like a "thing." It could disintegrate by noise from graininess in the potential. (It would almost certainly have disintegrated had the "whipping about" been numerical.) Why do the particles hang together so strongly? We can only explain it as a consequence of the self-consistency of the entire system. But the disk is only 1% of the mass, and the fraction of that mass down at the center is much less than that. If self-consistency is not important, we should be able to duplicate the effect by some independent particles in an external potential. But when we put a set of independent particles in a potential and follow the trajectory of the normal on the unit sphere, the trajectory soon loops the pole--something we do not see in Figure 1. The motion is different. Can we measure the importance of self-consistency for this system?

What is the role of self-consistency? Must the disk be self-consistent in order to remain as thin and as sheetlike as it does, for as long as it does, in spite of warps? Does self-consistency help the disk avoid becoming corrugated? Does self-consistency of the background galaxy play any part in the longevity of the disk? These questions might be tested by means of experiments in which only parts of the system are self-consistent. For example, we might replace the galaxy with a fixed potential while keeping the disk self-consistent. This would test the role of self-consistency in the disk for all the questions raised here. We might then shift the center of the fixed potential that represents the galaxy, to study whether such a shift induces motions like those seen in the experiments.

The wavelike patterns in Figures 4 and 6 suggest that nutation moves in wavelike patterns. Is this an effect of self-consistency in the disk, or is it simply a kinematic wave that arises because the original excitation was pretty much coherent and the frequencies differ little from place to place?

We suggested that the galaxy center may have moved. How difficult is it for a galaxy to move its center? There is no physical reason why a galaxy cannot decide that it would like to have its center somewhere else, but how hard is it to move the center? One measure of this would be to take a self-consistent galaxy model and move its center relative to its envelope. Then one could try to determine the oscillation frequency experimentally, if the system oscillates. A low frequency means that the center is easy to move. If you try to estimate the frequency, a first guess comes out that the mode may be neutral. If so, there would be no resistance to shifts of the galaxy center. We recently ran an experiment to test this. The galaxy model had its center shifted relative to its envelope. We cannot see any evidence that the galaxy center is oscillating within the galaxy envelope. It looks as if this mode has a very low frequency--that the density center is easy to shift away from the geometrical center.

We mentioned that it is difficult to describe the shape of the disk. One of the experimental questions we would like to answer is how much the disk fattened up from a warped sheet. But we cannot determine this without some way of determining where the (warped) disk is, so we can measure how far a particle finds itself away from that mean disk.

5. DISCUSSION AND FUTURE PLANS

The longer run, AWOBL5, shows that a disk can survive for quite a long time in an oblate galaxy. It warps somewhat, but it doesn't become as convoluted as theoreticians had feared (Toomre 1983, Sparke 1984). Our experiments scale to galaxies by taking a crossing time as 50 million years (about one radian of the sun's 250-million-year orbit about the Galactic center), so 22 crossing times for AWOBL5 is equivalent to 10^9 years. This is getting on toward a reasonable fraction of the age of a galaxy. Even if these dark-lane ellipticals are transient events, as is often supposed (Wilkinson *et al.* 1985), they last long enough that there seems to be little difficulty in accounting for the number known in terms of an embedded disk.

The dynamics of the disk in a galaxy is certainly rich enough, and the disk is sufficiently long-lived, to suggest that this model is promising as a means to study polar ring S0's and dark-lane elliptical galaxies. This leaves us with the challenge to see how such a condition might have been established in the first place. The model is on a solid footing for the study of disk galaxies in massive haloes.

We plan to extend this work to include triaxial systems, generalizing beyond the oblate symmetry employed so far. Motions and responses are likely to be even more complicated in that case. The richness encountered in the simpler case tells us that there is a lot to study and a lot more to learn before we tackle that more complicated situation.

The extraordinary motions of the disk center in these experiments reminds you of precessing jet sources and of the complicated Galactic center region as revealed by the observations (see, e.g., Brown and Liszt 1984). Those observations show strange velocity fields and even stranger shapes. The velocity fields have been mapped in various IR and radio spectral lines that testify to unusual densities and unusual ionization conditions. Conventional arguments place a massive black hole at the Galactic center, and invoke motions in its neighborhood along with accretion phenomena, etc., to explain what is observed. However, motions like those seen in the experiments could account for many of these observations if something like those motions is going on at the Galactic center. Giant molecular clouds banging into other giant molecular clouds, as each gets swept along in these orbiting motions, could produce observations showing large velocity differences across small projected distances. They could lead to strange excitation and ionization conditions in shocks where two colliding clouds intersect. The problem then is to determine whether motions like this could go on at the center of a galaxy, and whether they are going on in our Galaxy. The out-of-equilibrium starting conditions doubtless helped start these motions in the experiments, and an important question is whether the center could move in our Galaxy or in other galaxies. This leads back to the puzzle how hard it is to make a galaxy's center shift. The preliminary experimental result mentioned earlier, that the shift produces very small energy changes and acts like a neutral mode (or at least like a mode with unusually low frequency) indicates that the center could move easily in response to a stimulus, or that it might even move spontaneously. I need not remind this audience that disturbances characterized by spherical harmonics with $l=1$ are extremely difficult to excite by means of gravitational forces.

On more immediate grounds, the physical origin of the shift of a galaxy's center is a challenging topic which we are pursuing at the moment. Our main problem, as has always been the case with our projects, is that there are so many exciting questions and ideas crying for attention that we have far more to work on than we can possibly manage.

It is a pleasure to acknowledge the stimulating teamwork of Bruce Smith and Althea Wilkinson. Most of this study has been carried out at the NASA-Ames Research Center, and funds for partial support of this study have been allocated by the NASA-Ames Research Center, Moffett Field, California, under Cooperative Agreement NCC 2-265.

REFERENCES

Brown, R. L., and Liszt, H. S. 1984, Annual Reviews of Astronomy and Astrophysics, **22**, 223.

Kotanyi, C. G., and Ekers, R. D. 1979, Astronomy & Astrophysics, **73**, L1.

Miller, R. H. 1978, Astrophysical Journal, **223**, 122, Fiche 67-C12.

Rubin, V. C., Burstein, D., Ford, W. K., and Thonnard, N. 1985, Astrophysical Journal, **289**, 81, Fiche 13-G3.

Sparke, L. S. 1984, Astrophysical Journal, **280**, 117, Fiche 50-C8.

Toomre, A., 1983, in Proceedings of IAU Symposium 100, Internal Kinematics and Dynamics of Galaxies, ed. E. Athanassoula (Dordrecht: Reidel) pp. 177-186.

Tubbs, A. D. 1980, Astrophysical Journal, **241**, 969, Fiche 110-C7.

van Albada, T. S., Kotanyi, C. G., and Schwarzschild, M. 1982, Monthly Notices Roy. Astr. Soc., **198**, 303.

Wilkinson, A., Sharples, R. M., Fosbury, R. A. E., and Wallace, P. T., 1985, preprint, STELLAR DYNAMICS OF CEN A.

A REVIEW OF ASTROMETRIC OBSERVING

R. S. HARRINGTON
U. S. Naval Observatory
Washington, DC USA

A review of all aspects of astrometric observing, if it is not to turn into a textbook, can consist of little more than a list of all types and techniques of such observing, with a brief summary of the associated applications. Indeed, Kovalevsky (1984) has recently given such a review as part of a more extensive discussion of the Hipparcos/Tycho project, and the present review will borrow heavily from that one. However, specialized topics such as "experimental" astrometry (radar ranging, spacecraft tracking, etc.) or occultation astrometry will not be considered. Speculative astrometry, such as observing with the Hubble Telescope or dedicated satellites also will not be considered.

Kovalevsky breaks groundbased astrometry down into five basic types, depending on the sizes of the arcs being measured, or equivalently, the area of the sky being considered in a given set of observations. That same classification scheme will be followed here, although the order will be reversed. For each of the five types of astrometry, something can be said about the observing instrument, the detector, and the applications of the results.

The most fundamental type of astrometry is GLOBAL. That is, an observational realization of the entire sky, in such a way that a truly fundamental reference system is obtained, is the primary objective of modern astrometry. The area of sky under consideration is the entire 41253 square degrees, and that, unfortunately, is not obtainable from the ground. The uninitiated might think a transit circle on the equator could do the job, but it is impossible to locate the poles, and hence establish a fundamental system, without circumpolar stars. Hence, present efforts in this type of astrometry are limited to computers applied to catalogs of the next type of astrometry, but there will be a qualifier shortly.

The next type of astrometry could be called HEMISPHERIC, or quasi-global. This covers all of the sky obtainable from a given observatory, with the proviso that the observatory is far enough from the equator that it can locate a pole, but not so far that it loses the ecliptic. The area of the sky covered is typically of the order of 30000 square degrees, or a substantial part of the sky.

Classically, the instruments normally used in this area are transit instruments and meridian circles, or nowadays the two-for-one transit circles almost exclusively. However, astrolabes are still employed in

Celestial Mechanics **37** (1985) 325–328.

this field, and it must be remembered that photographic zenith tubes, while philosophically in reverse gear, are still basically carrying out this type of astrometry. A new type of instrument has now appeared - the interferometer. The radio interferometer has already proved to be probably the most accurate astrometric instrument available, and it now promises to provide the ultimate link to the almost inertial reference system. Experiments with infrared and optical interferometers are also already producing promising results.

The detector for the traditional instruments has been the human eye, peering through a micrometer of some sort. This has had the disadvantage that it has put a warm body in the observing environment, and its effects are quite detectable at modern-day astrometric precisions. For some instruments it has been possible to substitute the photographic plate as the detector. For the transit circles, the trend is to place some sort of photoelectric device feeding a computer in place of the traditional micrometer, removing the observer to the warmth of the computer room.

The ultimate application of hemispheric astrometry is to produce a catalog that can be used to establish a true fundamental reference system. If the angles measured with these instruments are measured with respect to fundamental points (e.g. a pole and an equinox), then such a result is possible, and such catalogs as the FK series are indeed compilations of such efforts. It is also possible to obtain positions (and motions if patient enough) differentially with this type of effort, but such positions, while absolute in some sense, are not fundamental. Again, reversing the philosophy, by assuming a reference system it is possible to determine time and geodetic information, and sometimes a bit of everything gets into the picture.

There is a way to use this type of astrometry to almost achieve a first-type situation, and this is the approach being presently pursued at the U. S. Naval Observatory. That is to have a transit circle in each hemisphere, observing simultaneously, with a substantial overlap region of the sky. Each instrument establishes its own system, but by imposing certain obvious boundary conditions, such as the North and South Poles are 180 degrees apart and the stars' positions are telescope-independent (the diurnal parallax of Alpha Centauri is only 40 micro-arc-seconds!), a quasi observational realization of the entire sky is possible.

The next type of astrometry is WIDE FIELD. The area of the sky covered in a given field may be 20 to 50 square degrees, and this is the field of the astrographic cameras. Traditionally, the instrument has been the medium speed astrographic telescope, but recent years have seen the high-speed Schmidt camera enter this field. Although not originally designed to do astrometry, Schmidts can make a significant contribution with appropriate care and prudent computing. All of this work is photographic, and it will remain so for some time to come. The results are right ascensions and declinations (and motions with time), but these are relative and hence not fundamental. Differential astrometry, using catalogs that are themselves differentially derived, can be carried out to obtain positions of specific objects, such as planets, satellites, or comets. It is also possible, by imposing the constraints of an overlap scheme, to obtain a meaningful approximation to an entire sky realization. However, it must be remembered that catalogs such as the SAO or the AGK3 are basically compilations of wide-field results.

Next down in size comes NARROW FIELD astrometry. Here are fields of view typically less than, although a substantial fraction of, a square degree. This has been the domain of the long-focus, sub-meter-aperture refractor, complete with all of the potential difficulties associated with such large pieces of glass. There have been modern improvements in this field with the 1.55-meter astrometric reflector of the Naval Observatory, its 2/3rds scale clone in Torino, and the 1-meter reflector of McCormick Observatory. In addition, a 1.56-meter astrometric reflector is nearing completion in China. It is also possible to do this type of astrometry with more conventional reflectors in some cases, but great care is required.

The detectors that have been used to date have been almost exclusively cameras, but, because they are so inefficient and non-linear, they will almost entirely disappear well before the end of the century. Several groups are developing various types of photoelectric detectors, but the real promise seems to lie in the field of charge converters (CCD arrays, CID arrays, etc.).

It is possible to obtain global positions with this type of astrometry, and indeed this is necessary to relate any optical reference system to something that is almost assuredly inertial. This requires a multiple differential process, a process generally called cascading when it reaches this realm. However, this field has been primarily devoted to obtaining differential parallaxes and proper motions, and indeed this is still the only way to determine astronomical distances directly. It is also possible to examine multiple star orbital motion by this technique and hence determine stellar masses. If orbital motion is detected without a companion, the existence of the companion can then be deduced, and it is upon this approach the principal hopes for extra-solar planetary detection lie.

The final type of astrometry can, by comparison, be called NEGLIGIBLE FIELD. The areas involved are 100 square arc seconds or less (a few micro square degrees), and the telescopes are the same as those used for narrow field astrometry. Photographic cameras can be used for detectors in such systems, but they are severely limited in that photographic emulsion resolution is generally much worse than large aperture resolution. For that reason, the eye at the micrometer is still a usable detector and is still used in some instances. Charge converters or other electronic devices have a future here, and a particular application deserves special mention. The speckle interferometer takes advantage of the short-period atmospheric stabilities to freeze the pattern of speckles produced in a very short exposure and then combine the "specklegrams" to obtain resolutions approaching the theoretical limits of large telescopes. The applications here involve only relative positions, either of a satellite with respect to a planet or another satellite, or of components of multiple star systems. The objective is to obtain relative orbits, and hence whatever these can teach, such as masses, figures, etc.

A recent book reviewer, who will remain anonymous, characterized "painstakingly precise measures of the positions and motions of celestial objects" as "hidebound, hoary, and just plain dull". If that is the case, the variety of ways now available to be "just plain dull" is a testament to human ingenuity, and the variety of applications indicates that the field is absolutely vital to astronomy in general and in many cases makes quite exciting contributions to it.

REFERENCE

Kovalevsky, J.: 1984, Space Sci. Revs. 39, 1.

Precise Position Measurements of Jupiter, Saturn and Uranus Systems with the Very Large Array

D.O. Muhleman, G.L. Berge, D.J. Rudy
California Institute of Technology
A.E. Niell, R.P. Linfield, E.M. Standish
Jet Propulsion Laboratory

1. INTRODUCTION

The discovery of Quasi-Stellar Objects (QSO's) and the development of Very Long Baseline Interferometry (VLBI) techniques since the early 1960's have afforded an extremely precise coordinate grid system or reference frame on the sky. Since the QSO's are both stellar-like objects and strong emitters of radio flux, their positions can be measured with both optical and radio techniques. However, the radio VLBI measurements on simple, compact sources are by far the most precise, the useful unit being micro-radians. The origin of the VLBI reference frame is tied to the stellar reference frame by the measurements of Hazard et al. (1971) who precisely measured the optical position of the QSO 3C273 with lunar occultations, which, when combined with VLBI position measurements of the same source, ''tied'' the VLBI and stellar position frames together. Their reported accuracy is ±0.15 arc seconds, but there is good reason to believe that the error in that frame tie is significantly smaller. Radio tracking measurements of spacecraft associated with the inner planets, such as the Viking Orbiters around Mars, combined with VLBI measurements indicate that the frames of the inner planets (i.e. their ephemerides), stellar system, and the VLBI reference sources are consistent to better than $\pm 0\overset{''}{.}03$ (Newhall et al., 1984).

In the work reported here, we have attempted to utilize the outer planets for the same purpose with the additional goal of improving the outer-planet ephemerides. As yet no spacecraft have been put into orbit at any major planet and the ephemerides are essentially based on the traditional optical astrometric data types. However, Jupiter and Saturn do have large natural satellites which emit sufficient radio flux for position measurements and Uranus is sufficiently small in angular extent

Celestial Mechanics **37** (1985) 329–337.

that its position can be accurately measured with radio-interferometry techniques. In this paper, we are reporting the results of an extensive campaign to measure accurate positions of these planets. None of this work is yet published. The measurements of the Jupiter system were made by G. Berge, A. Niell, and D. Muhleman; those of the Saturn system by G. Berge and D. Muhleman; and those of Uranus by G. Berge, R. Linfield, and D. Muhleman. Measurements of Neptune are planned for 1986. Most of the measurements discussed here were made with the Very Large Array (VLA) of the National Radio Astronomy Observatory near Socorro, New Mexico operating at a wavelength of 2 cm (15 GHz). The results for the three planets are very similar and suggest that the ephemeris reference frame for the outer planets is in error by roughly -0.2 arc sec in right ascension. The details are discussed below.

2. MEASUREMENT TECHNIQUE

The VLA can be thought of as a collection of 2-element, phase coherent interferometers. If there are n antennas in the array there are n(n-1)/2 such interferometers operating. Usually for the VLA n = 27 and the instrument can be considered as a net of 351 such 2-element interferometers. The measurement principles can be understood by considering one such interferometer as shown in Figure 1. First, consider a point source which is emitting broadband radio noise (due to thermal emission in the case of a planet). Even though the radiation is noise, the flux is coherent across the phase front coming from the source which can be considered to be at infinite distance. We consider the problem in 2-dimensions in the diagram where the 2 antennas are fixed to the earth and separated by the baseline vector **B**. If the receivers at the 2 antennas are maintained in phase coherence by connection to the same reference oscillator, the signal arriving at antenna 2 is identical to that at antenna 1 but is delayed by $(B/c)\sin\theta$ seconds where c is the speed of light. A measurement of this delay yields the angular position θ of the source. In practice, the measurements are much more complex due to such effects as electronic phase variations in the 2 receivers, the lack of an absolute reference, delay variations in the Earth's atmosphere, intrinsic errors in measuring weak signals, etc. Most of these effects can be minimized if the source whose position is to be measured is within a few degrees on the sky of reference sources, in our case the QSO's. The error sources can be thought of as causing an instrumental phase error which is a function of time for a given antenna pair. This function is measured by periodically observing one or two QSO's near the planet. The position of the useful QSO's are known within the VLBI-QSO position network to an accuracy of about $\pm 0\overset{''}{.}02$ (Fanselow et al., 1984; Niell et al., 1984). The measurement of the planet's position is achieved by ''pointing'' each 2-element interferometer to the ephemeris position of the planet (see Fig. 1) with a continuous adjustment of the interferometer's phase by the VLA computer

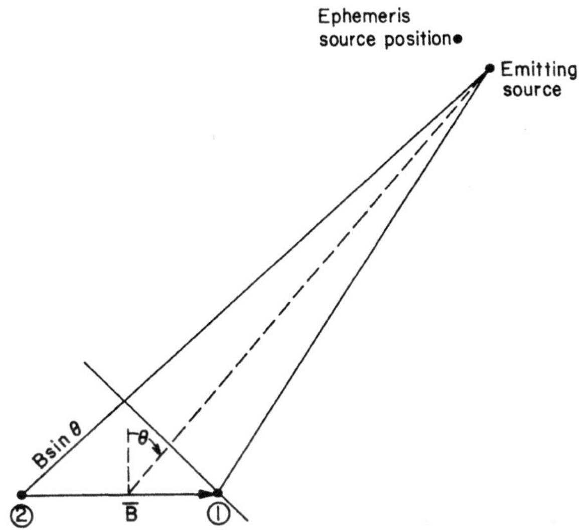

Figure 1. Basic geometry of an interferometry.

system. In all cases considered here, the planetary ephemerides were the JPL-Development Ephemeris 200. The positions of the QSO reference sources were taken from the JPL-VBLI position catalog (Fanselow et al., 1984; Niell et al., 1984). All of the necessary precessions and nutations were done by the group at JPL consisting of E.M. Standish, J. Leiske, A. Niell, and R. Linfield. In this way, the positions of the reference sources, the planets, and their satellites are independent of the VLA software (except for the diurnal aberration corrections) and are the total responsibility of the experimenters.

If the target planet or satellite is a point source (they are not), the response of our ideal, 2-dimensional interferometer depicted in Figure 1 would be

$$S = F \exp\left(\frac{2\pi i B}{\lambda} \sin \theta(t) \, \Delta\theta \right) \qquad (1)$$

where F is the flux density of the source averaged over the receivers' bandwidths, λ is the wavelength, $\theta(t)$ the angle shown in Figure 1 which varies as the earth rotates and $\Delta\theta$ is the ephemeris offset from the true-planet position (assumed to be very small). The visibility function, eq. (1), at an instant t is a complex number whose amplitude is the source flux density, F, and whose phase

$$\phi = \frac{2\pi B}{\lambda} \sin \theta(t) \, \Delta\theta \qquad (2)$$

contains the ephemeris offset, $\Delta\theta$. The generalization to 3-dimensions is straightforward.

The Jovian satellites and Titan exhibit highly symmetrical, circular disk brightness distributions on the sky. Uranus is very nearly the same except for a small asymmetry due to the orientation of the planet's pole and an associated latitude effect in the brightness. Using the approximation that these disks are uniformly bright, the observed visibility at time t is

$$S(t) = F_o \frac{J_1(2\pi\beta(t))}{\pi\beta(t)} \exp(2\pi i\beta\Delta\theta) \qquad (3)$$

where F_o is the total flux density emitted by the planet or satellite and

$$\beta(t) = \frac{B(t)}{\lambda} \frac{R}{D} \qquad (4)$$

where B(t) is the component of the baseline vector (for the specific antenna pair) perpendicular to the line of sight to the source, R the effective radius of the source, and D, the distance to the source. J_1 is the Bessel function of the first kind, of order 1. Equation (3) shows that the source is resolved by a specific interferometer as β increases since $J_1(x)/x$ asymptotically approaches zero as $x \to \infty$. Since the sources considered in this paper have disk sizes of order 1'', the B-array of the VLA at λ = 2 cm is nearly optimum for measurements of their positions.

3. OBSERVATIONS AND DATA ANALYSIS

The Jovian satellites: Europa, Ganymede, and Callisto were observed on 26 April 1983 and on 1 May 1983 within $\pm$3 hours of the VLA meridian. At that time, Jupiter was bracketed by 2 cataloged QSO's: NRAO530 and P1519-273 as shown in Figure 2. The satellites and the calibrators were observed in a repeated sequence. The instrumental phase function was determined by fitting eq. (1) to the QSO observations. After removing these phase functions from the satellite data, eq. (3) was fit to the data in a least-squares computation to determine the offsets of the true source positions of the satellites from JPL-DE200. Observations of planetary satellites must be carried out when the objects are near elongation relative to the central planet in order to minimize the confusion flux from the planet. Since the interferometer pairs are all phased to track the satellite, most of the flux from the central body is eliminated by interference effects when the data are combined over 6-hours. The geometry for the Jovian observations is shown in Figure 3. The satellite epherimides are from the work of J. Leiske and they were merged with the Jupiter ephermeris by E.M. Standish.

The results of the position determinations for Ganymede and Callisto are shown in Figure 4. The measurements of Ganymede are tightly clustered for the 2 days but those for Callisto show an obvious error in the orbit for that satellite which can be expressed as a time error along the orbit of 56 seconds. A relatively better orbit for Ganymede is to be expected since its orbital period is much smaller than Callisto's yielding more frequent and more precise observed eclipses of the satellite. The latter are a fundamental data source for the determinations of the satellite orbits.

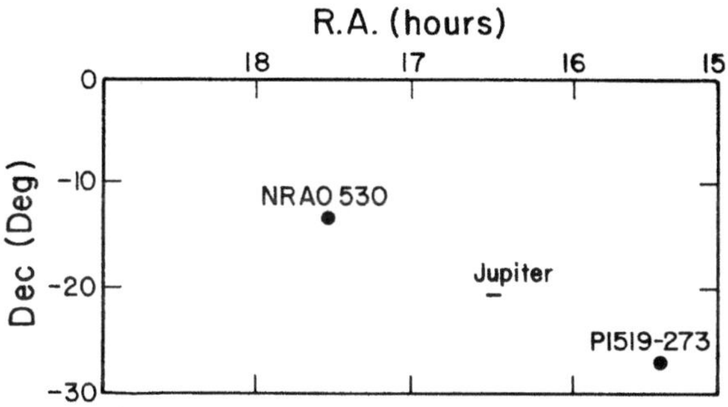

Figure 2. Positions of Jupiter and calibrators.

Titan was observed on 16 January 1984 with geometry as shown in Figure 5. Figures 3 and 5 show the extent of antenna beams (full widths at half maxima) for the individual VLA telescopes. It can be seen that the observations of Titan are considerably more difficult than those of Ganymede and, particularly, Callisto. The calibrator for the Titan observations was 1510-089. The offsets of the true positions of Titan relative to JPL-DE200 are shown in Figure 6. Observations were made at wavelengths of 2 cm and 1.3 cm. However, the receivers at 1.3 cm are much less sensitive than those at 2 cm. The latter results will be used to estimate the ephermeris offset of Saturn.

Uranus was observed at 2 cm with the VLA on 30 April 1985. The planet was observed in a sequence with two QSO calibrators: OT-111, Uranus, P1657-261, Uranus, ··· within ± 3 hours of its meridian crossing at the VLA. The results of a least-squares fit to the visibility data after the removal of the instrumental phase function are shown in Figure 7. The black dot shows the formal result of the fit. The visibilities were also Fourier transformed to form a map of the Uranus' flux distribution on the sky. This map shows the slight north/south asymmetry in the brightness distribution. This asymmetry was used to correct the least-squares position offset. The result is shown by the plus symbol in Figure 7. The position of the planet's centroid was also determined by measuring the position of the limb contours on the map and is shown in Figure 7 by the open circle. The 3 determinations are in close agreement but the declination correction for the brightness distribution is significant. The formal noise error on the least-squares determination is about $\pm 0\farcs025$ in both coordinates.

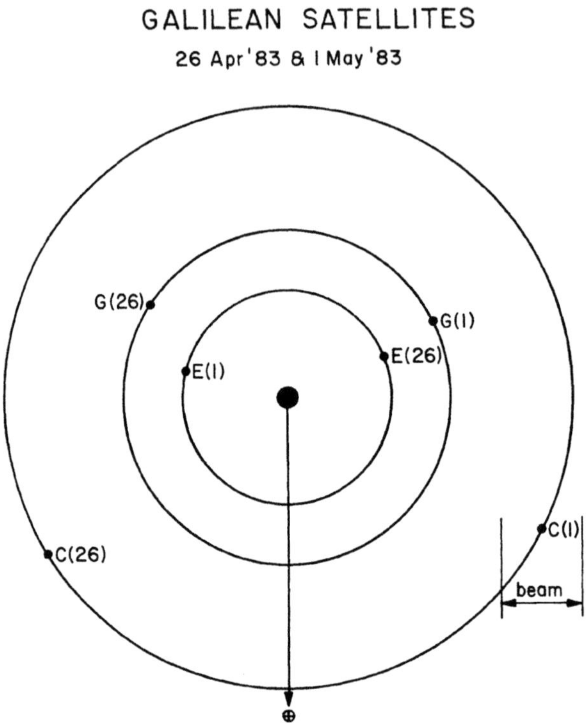

Figure 3. Geometry of the Jovian Satellite system.

4. RESULTS

Our best estimates of the ephemeris offsets for the major planets in the sense of true position minus JPL-DE200 position are:

Jupiter: 26 April - 1 May 1983 ($\alpha = 16^h30^m$, $\delta = -20\fdg9$)

$$\Delta\alpha \cos\delta = -0\farcs185 \pm 0\farcs025$$

$$\Delta\delta = -0\farcs058 \pm 0\farcs025$$

Saturn: 1617 January 1984 ($\alpha = 14^h53^m$, $\delta = -14\fdg2$)

$$\Delta\alpha \cos\delta = -0\farcs20 \pm 0\farcs03$$

$$\Delta\delta = +0.03 \pm 0\farcs03$$

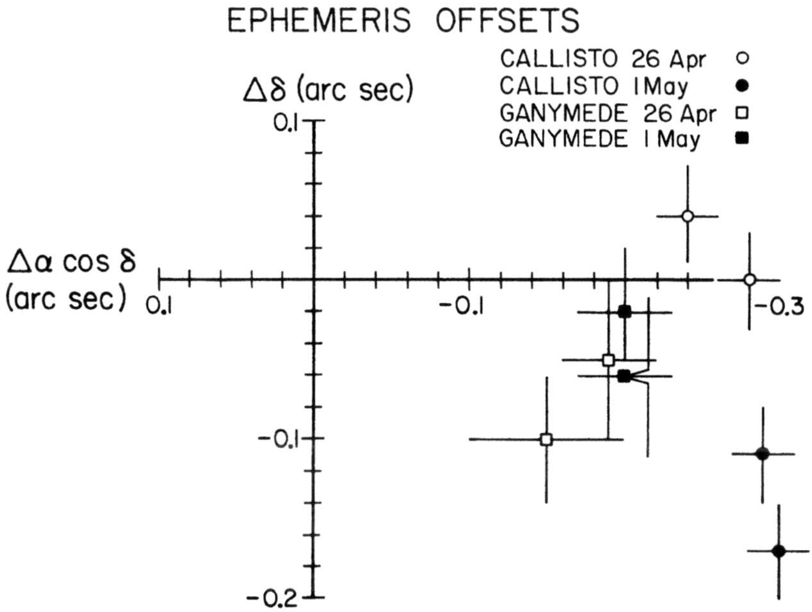

Figure 4. Ephemeris offsets of the Jovian system for 26 Apr and 1 May, 1983.

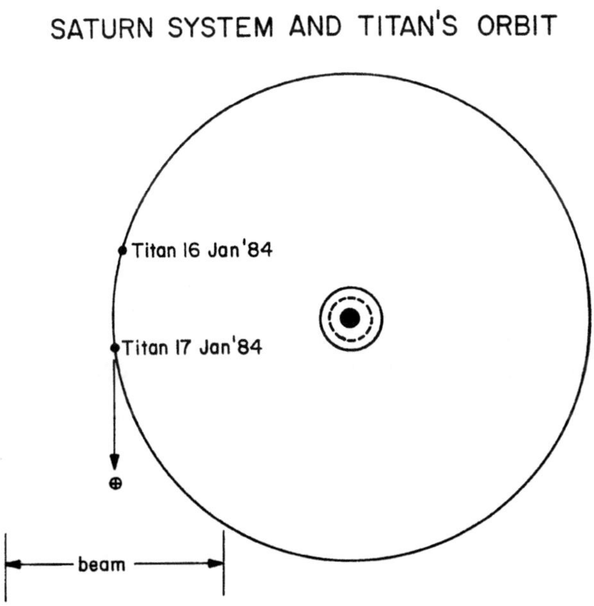

Figure 5. Geometry of the Saturn/Titan system.

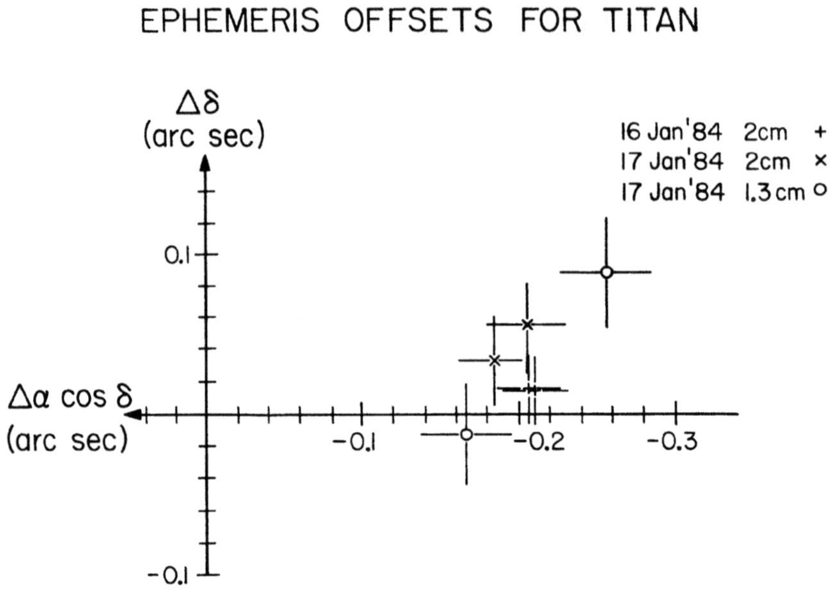

Figure 6. Ephemeris offsets for the Saturn/Titan system for 16, 17 January, 1984.

$$\text{Uranus:}\quad 30 \text{ April } 1985 \ (\alpha = 17^h 05^m,\ \delta = -22\overset{\circ}{.}9)$$

$$\Delta\alpha \cos\delta = -0\overset{''}{.}399 \pm 0\overset{''}{.}030$$

$$\Delta\delta = -0\overset{''}{.}041 \pm 0\overset{''}{.}035$$

All errors are formal estimates based on the noise in the data. The offsets for the three planets are very similar. We believe that most of the right ascension error is due to errors in JPL-DE200. It is premature to estimate what part of the offsets is caused by the reference frame tie between the stellar position system and the QSO's position system.

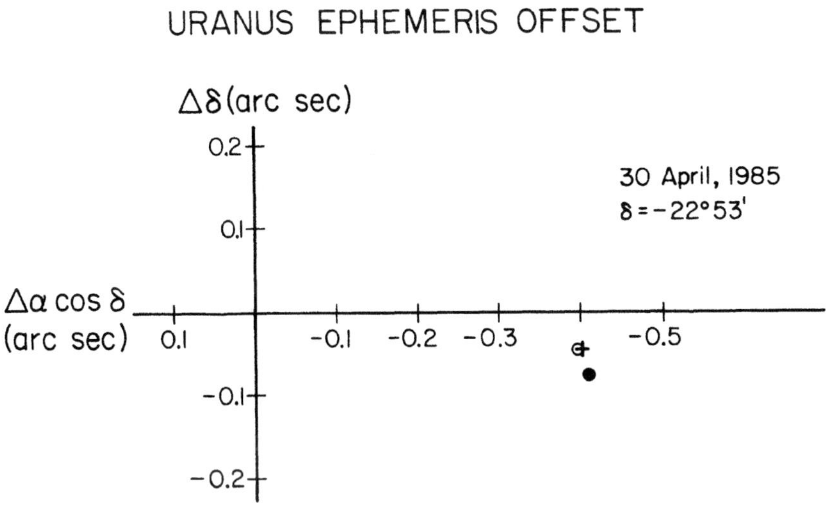

Figure 7. Ephemeris offsets of Uranus for 30 April, 1985. Filled circle is the least squares fit for the center of the flux density; the plus symbol indicates the position after correction for the pole position; and the open circle indicates the position based on the limb contours of the Uranus map.

5. ACKNOWLEDGMENT

We wish to thank the staff of the National Radio Astronomy Observatory for their assistance at the Very Large Array. D.O.M. was partly supported by the Jet Propulsion Laboratory for this work. The work at Caltech was partly supported by NSF AST 82-141790 and the President's Fund.

REFERENCES

Fanselow, J.L., Sovers, O.J., Thomas, J.B., Purcell, G.H., Cohen, E.J., Rogstad, D.H., Skjerve, L.J., and Spitzmesser, D.J. Radio interferometric determination of source positions utilizing Deep Space Network antennas - 1971 to 1980. A. J., **89**, 987-998, 1984.

Hazard, C., Sutton, J., Argue, A.N., Kenworthy, C.M., Morrison, L.V., and Murray, C.A. Accurate radio and optical positions of 3C273B. Nature, **233**, 89-91, 1971.

Newhall, X.X., Preston, R.A., and Esposito, P.B. Relating the JPL VLBI reference frame and the planetary ephemerides. Proceedings IAU Symposium No. 109: Astrometric Technique, Gainesville, Florida, 1984.

Niell, A.E., Fanselow, J.L., Sovers, O.J., Thomas, J.B., Liewer, K.M., Treuhaft, R.N., and Wallace, K.S. Accurate positions of 120 radio sources with declinations above -45 degrees. Proceedings IAU Symposium No. 109: Astrometric Techniques, Gainesville, Florida, 1984.

COORDINATION OF MINOR PLANET ORBITAL STUDIES: PAST, PRESENT AND FUTURE *

Brian G. Marsden
Harvard-Smithsonian Center for Astrophysics
Cambridge, MA 02138
U.S.A.

ABSTRACT. A brief account is given of the coordination of astrometric observations and orbital studies of minor planets, particularly with regard to the role played in the past by the Astronomisches Rechen-Institut. The current situation is surveyed and thoughts are presented on future requirements.

The Astronomisches Rechen-Institut was established in Berlin in 1874, largely on account of a pressing need to organize the computation of ephemerides for the Kleine Planeten, or minor planets. As is well known, the first four of these objects were discovered over the course of the first seven years of the nineteenth century, and after a hiatus of almost four decades, new discoveries were being made with increasing regularity. As director of the Berlin Observatory and editor of the _Berliner Astronomischer Jahrbuch_ until his death in 1865, J. F. Encke had overseen the ephemeris computations for many years; but in the early 1870s, by which time some 120 Kleine Planeten were known, Encke's successor, W. Foerster, was strongly urging that another institution, distinct from the Berlin Observatory, should be established for attending to the minor-planet work, as well as to the other calculations involved in the production of the _Jahrbuch_.

It is important to draw a distinction between the work on Kleine Planeten carried out under the initial direction of F. Tietjen at the Astronomisches Rechen-Institut and that handled by the editor of the _Astronomische Nachrichten_, at the time C. A. F. Peters and soon afterward A. Krüger, in Kiel. The _A.N_. was responsible for the announcement of new discoveries and the publication of early unperturbed orbits and ephemerides. Although many of these early orbits and ephemerides were calculated at the A.R.I., the A.R.I. was principally concerned with what was then the immensely more troublesome problem of

* Presented at "Frickefest", a colloquium held in Austin, Texas, on March 27, 1985, in honor of Walter Fricke on the occasion of his retirement as director of the Astronomisches Rechen-Institut in Heidelberg.

Celestial Mechanics **37** (1985) 339–344.

incorporating perturbations into the computation of ephemerides at future oppositions of the growing collection of objects. This in effect required the preparation of a general theory for the motion of each new body.

In 1896 J. Bauschinger succeeded Tietjen as A.R.I. director, and H. Kreutz became A.N. editor. By then the photographic surveys by M. Wolf at Heidelberg and A. Charlois at Nice had almost entirely displaced the traditional visual searches for minor planets. Discoveries, or suspected discoveries, were initially being designated by letters, and minor planets were not numbered until an orbit had been calculated. Thus the famous discovery "DQ" in 1898 became number (433) and also received the name Eros. Orbit calculations were not always particularly reliable, and many of the numbered objects were promptly lost. The perturbation problem was greatly alleviated following a suggestion by K. Bohlin that approximate but adequate Hansen-type perturbations could be produced for whole groups of minor planets with similar mean motions.

The editorship of the A.N. was transferred to H. Kobold in 1907, and a year or so later F. Cohn succeeded Bauschinger as A.R.I. director. Sept. 1907 also brought planet "ZZ", discovered in Heidelberg by A. Kopff, and this was then followed with a new "AA". These two objects became (643) Scheherezade and (644) Cosima, and in 1924 discoverer Kopff became A.R.I. director. By then A.R.I.'s responsibilities were increasing in other areas, most notably in the production of the fundamental star catalogues, and the actual responsibility for the Kleine Planeten devolved on G. Stracke. It had been Cohn's aim that the ephemerides of the minor planets should be accurate to 30-60 arcmin, and to a commendable extent he met this aim. The annual batches of ephemerides of minor planets were separated from the Berliner Astronomischer Jahrbuch into the annual publication Kleine Planeten, and discovery announcements were removed from the A.N. itself to its series of Beobachtungs Zirkulare. When Stracke took over the reins about 1000 Kleine Planeten had been numbered, and the alphabetic system of provisional designations was recast in its present form, which allows more satisfactorily for the interpolation of objects belatedly recognized on older photographic plates. The series of Rechen-Institut Zirkulare was also initiated. Although the "R.I.s" had a very small distribution and were initially mimeographed from hand-written sheets, they gave detailed immediate data on minor planets and for workers in the field soon became more authoritative than the well-distributed, properly typeset, B.Z.'s. Unfortunately, poor print quality and the high sulphuric-acid content of the paper has made preservation of the R.I.'s a severe problem; since the R.I.'s are by far the principal source of observations during the 1930s, a decade of unprecedented growth, the potential loss has been all the more grave.

Although the devastation of World War II resulted in the collapse of the A.R.I. in Berlin it was subsequently revived as two separate A.R.I.'s, one in Heidelberg (with Kopff again as director, until he was succeeded by W. Fricke in 1955), the other under the direction of A. Kahrstedt in East Berlin. But the devastation, coupled with the death of Stracke in 1943, brought termination to both the R.I.'s and the annual Kleine Planeten volumes; most of the stock of the 1945 volume was destroyed at the printers, and only a few copies of the manuscript for 1946 and of the last dozen of the 2569 R.I.'s were ever distributed. A total of 1564 minor planets had been numbered by this time, but well over 10 percent had orbits that came nowhere close to meeting Cohn's accuracy criterion. When some semblance of normalcy resumed, work on minor planets was resumed under the leadership of E. Rabe and later W. Strobel in Heidelberg and of Kahrstedt himself in East Berlin. The East Berlin effort lasted only a few years and was mainly restricted to computations on earth-approaching objects, while the more extensive activity in Heidelberg was included in, for example, the new A.R.I. series Nachrichtenblatt der Astronomischen Zentralstelle.

The future international coordination of orbital studies was, however, to be from elsewhere. In 1947, the International Astronomical Union--through Commission 20--established the Minor Planet Center at the Cincinnati Observatory under the direction of P. Herget, assisted by Rabe on his transfer from Heidelberg in 1948. The Minor Planet Circulars, issued by the Center, were to supersede the R.I.'s. Fortunately, Herget was immediately able to bring in automated methods for the computation of special--rather than general--perturbations, although for many years other aspects of the work were still done by hand; some of the orbital differential corrections were in fact done by A.R.I. staff members in Heidelberg in collaboration with Herget and Rabe. As for the observations that were a necessary prerequisite to the differential corrections, Herget was able to interest F. K. Edmondson in getting students at Indiana University involved in an observing program using the Cincinnati 25-cm refractor, which was moved to the Goethe Link Observatory for this purpose. The successor to the Kleine Planeten volumes was Efemeridy Malkyh Planet, prepared at the Institute for Theoretical Astronomy in Leningrad under the direction of M. F. Subbotin; about half of the 1947 opposition ephemerides were published by the Watson Computing Laboratory in New York, but from 1948 onward I.T.A. prepared the complete volume.

Because of the great need to improve the quality of the orbital data for the existing set of numbered minor planets, Herget drastically reduced the rate at which new objects were numbered. To qualify for numbering a minor planet generally had to have been reasonably well observed at three or more oppositions--or perhaps only two for well-observed objects that could come close to the earth. This caution was clearly desirable, and it was not until the early 1970s that new objects were again being numbered at the rate of the 1930s. When he retired in 1978, and the Minor Planet Center moved to the Smithsonian Astrophysical Observatory, Herget had added fewer than 500 new objects to the numbered list, and of the few dozen objects with ephemerides in error by more than 10 arcmin, just 23 were low-numbered planets lost because there were no clearly demonstrable observations at more than one opposition. Much of this great improvement in accuracy has of course been due also to the work of the Minor Planets and Comets group at I.T.A. This was led by S. G. Makover until shortly before his death in 1970 and has since been the responsibility of Yu. V. Batrakov. The more recent Heidelberg A.R.I. activity has involved both orbit improvements and more theoretical studies and has been an enterprise, not of a single organized group, but of individuals such as J. Schubart and L. D. Schmadel. A significant improvement, beginning with the 1980 volume of _Efemeridy_, has been the tabulation of all the orbital elements for a single epoch of osculation during the year and consistent with subarcsec precision in ephemerides. Only some 1 percent of the 3226 numbered minor planets now have ephemerides that are in error by more than 1 arcmin; six of these objects are lost, a situation that has remained stable now for some three years.

After the Indiana observing program ceased operation in 1966 the bulk of the observations were being obtained in an I.T.A.-sponsored program at the Crimean Astrophysical Observatory. With its limiting magnitude of B = 18 the Crimean program essentially sets the standard for observing programs that yield a fair number of unidentified minor planets. Some observers also report fainter objects, but while identifications for them may sometimes be found, this tends to be an unnecessary exercise unless the observer is prepared deliberately to follow up a fair proportion of his discoveries for a couple of months. The Minor Planet Center issues magnetic tapes with its complete set of more than 300 000 observations; following a partial effort to salvage valuable material from the crumbling _R.I._'s, the tape is intended to include all observations of unnumbered minor planets, but to be complete for observations of the numbered minor planets back only to 1939. Many of these older observations were not accurately measured at the time, but a few observatories are now trying to rectify matters. Thought has been given to extending the machine-readable file of numbered planets to include more of the pre-1939 accurate positions. Unfortunately, many of the observations of the lower-numbered planets in the files are affected by gross errors. A project is underway to remedy the situation with the help of a century-long integration of orbits at A.R.I. Heidelberg by H. Scholl, but progress is rather slow.

The monthly batches of _Circulars_ purport to include suitable elliptical orbit computations for all new objects that are sufficiently observed. Several readers are currently using these orbits and the machine-readable file of observations of unnumbered minor planets to identify observations of the same object at different oppositions, and some of them assist the Minor Planet Center in calculating orbits that link these oppositions. The _Catalogue of Orbits of Unnumbered Minor Planets_, a publication of the Center in 1982, contains 2471 entries (but excludes the comparable number of orbits from the Palomar-Leiden Survey of 1960). This represents a significant increase over the 600 elliptical orbits in a similar publication (which also contained many circular orbits) in 1961, but the number of such elliptical orbits has now increased to as many as 4500. A companion volume, _Catalogue of Discoveries and Identifications of Minor Planets_, lists the observatory involved and shows a reference for each of the 30 000 discoveries given provisional designations. This was an update--in abbreviated form--of the old A.R.I. _Identifizierungsnachweis_ publications, the last of which had been published by Strobel in 1963. Many of the orbits can also be accessed by "dialing-in" over a modem to the computer service operated by the Minor Planet Center (in conjunction with the Central Bureau for

Astronomical Telegrams), and the service then automatically allows the user to generate ephemerides as required.

Electronic communication will clearly play an even greater role in the the coordination of minor-planet work in the future, and one can foresee the day when the printed Circulars and Efemeridy volumes no longer exist. Ideally the computer service would inform the user of the objects in greatest need of observation, and its output would directly set his telescope to observe them. Such special coverage might refer, for example, to objects shortly before their predicted occultations of relatively bright stars. It is clearly feasible, also, that the computer service should allow a user to identify for himself observations of numbered minor planets beyond the range of the standard opposition ephemerides, but the present cost algorithm would cause him to be charged as much as $6.00 per run. This will presumably change with the introduction of less expensive computers, and the future computers will also presumably further facilitate the process of finding identifications. Indications are that the need for more minor planets with reliable orbits will continue, principally so that one can explore in ever-increasing detail the structure of the belt of minor planets, particularly as it relates to the less obvious Hirayama families.

On the observational front, two or three observers have already been exploring, just during the past year, the use of the charge-coupled device as a replacement for the photographic plate, and while there are still sometimes difficulties in assuring adequate reductions of the resulting astrometric positions, this promises to have a revolutionary effect on the subject.

Finally, there is the matter of observational accuracy. At a colloquium honoring a man renowned for his activities in fundamental astrometry, it is embarrassing to confess that, for the most part, minor-planet astrometry is content to have an observational accuracy of 1-2 arcsec! Of course, there are special situations, where one would like to have observations that are an order of magnitude more accurate, but since the general coordination of minor-planet work necessarily involves observations of such an enormous number of objects, most of which are followed only briefly, it would be foolhardy to attack each observation with the special tools needed to achieve this greater accuracy. If observers want to put more effort into their data they could first ensure that the date and time (preferably to the nearest second) of each observation are properly recorded, for these are by far the most frequent sources of gross error! Special situations, such as the use of minor planets for mass determinations, occultations, radar and radio investigations, space-telescope callibration, and the like, are occurring with increasing frequency, but these involve less than 1 percent of the observational data. Minor-planet astrometry still utilizes the equinox B1950 and what is basically the FK4 system (through the intermediaries of the AGK3 or SAO Catalogues). At some time (when appropriate star catalogues are widely available) it would probably be reasonable to introduce J2000 and the FK5, but one needs to consider whether it is really desirable indiscriminately to apply the necessary transformations to the existing voluminous and heterogeneous set of observational data. If the day should ever arrive when it really becomes desirable to improve the accuracy of the existing data, one would hope that it follows the invention of a device capable of remeasuring and doing a global re-reduction of an old photographic plate in a matter of seconds!

FRICKE'S INFLUENCE ON THE WORLD OF ASTRONOMY

Gart Westerhout
U. S. Naval Observatory
34th and Massachusetts Avenue, NW
Washington, DC 20390

It gives me great pleasure to highlight some facets of the career of Walter Fricke. As I do, it becomes clear that by simply listing his written contributions one already gets a feel for the intensity with which this man has contributed to, and has influenced the field of astronomy. When reading some of the material he produced one cannot but stand in awe of his great didactic qualities.

Walter Fricke started out as a student in Berlin, working on spiral nebulae, redshifts and cosmology. His first paper, published in 1937, dealt with the effect of the redshift on the brightness of spiral galaxies. McVittie, an established astronomer in Great Britain, critized his first paper; this meant that his very first paper received notice. Characteristically, Fricke did not take that criticism lying down and, at the age of 23, very politely stated in an answer to McVittie in the Zeitschrift für Astrophysik that the learned gentlemen from England did not understand the data. He pointed out that one must always discuss the observations first. Interestingly enough after World War II McVittie and Fricke found that they had been working on different sides of the war in the same field: coding and decoding machinery, trying to decipher coded messages. His doctoral thesis indicated his interest in dynamics. He studied the effect of dust clouds on the dynamics of dense star systems and concluded that dust clouds do not do a thing to such systems; the dust is entirely pushed by the stars. The relaxation time of the dust is of the order of 10^3 years while that of the stars is more like 10^{10} years. This thesis was an excellent piece of dynamical and statistical astronomy. In it he did an ab initio calculation of relaxation time. At Hamburg, after World War II, Fricke started work on our own galaxy. He was especially interested in the observational and theoretical work on galactic rotation and did a detailed rediscussion of the high velocity stars. This interest led him to realize the need for absolute proper motions, as well as radial velocities. In fact, he wrote the chapter on Radial Velocities in the 1952 edition of Landolt-

Celestial Mechanics **37** (1985) 345–348.

Börnstein. He also got involved in the field of observational astronomy. He studied the Andromeda Nebula with the Hamburg Schmidt telescope and even designed, built and described a sensitometer. This work on the Andromeda Nebula led to several papers on the intensity and color distribution in the nebula. This aided him in writing a classic 25-page paper in 1952 on the Kinematics and Dynamics of the Milky Way. In this paper, he did a large amount of modeling and came to the conclusion that the mass distribution and thus the major kinematics in our galaxy are determined by the Population II objects while the cloud population and the spirals play only a secondary role.

It is interesting to note that in the first 10 years after World War II he published 20 papers of which at least six were review papers. The educator in Fricke was present from the start and many of us have profited greatly from his didactic abilities combined with his willingness to spend hours in discussion and explanation with all comers.

Fricke calls the highlight of his career the year he spent in the United States from 1953 to 1954. Fricke and Schluter were the first to get an unencumbered fellowship from the German Science Foundation allowing them to travel freely throughout the United States. He worked with Chandrasekhar on gravitational stability. He did photoelectric photometry with Hiltner at McDonald and worked with Baade at Palomar. On his way back he visited a dozen observatories. At that point, he was already the German expert in Galactic Dynamics and Statistics and his appointment as Director of the Rechen Institut in Heidelberg in 1955 was therefore well deserved.

He started a new chapter in his educational task, this time in fundamental astronomy: "The Astronomical Coordinate System," published in 1957. He excelled in this field. In that same year he was also a major participant in an IAU Symposium on Galactic Structure and Dynamics. He was called in as the German expert but he came away as the man who would make sure that the best fundamental reference system is available for understanding stellar motions. This was the time of the discussion on Oort's Constants and the distance to the Galactic Center, and it became quite clear that much work on the observational material was necessary. He took over the Apparent Places of Fundamental Stars in 1960 (and started the invaluable Astronomy and Astrophysics Abstracts in 1969). He became the President of Commission 4 of the IAU, on ephemerides (1958-1964, two periods) and was involved in the astronomical constants revisions with a number of papers. He argued for the adoption for the new astronomical constants system, however incomplete, in 1964. In 1963 Kopff and Fricke published the FK 4. Younger people, who do not quite know the history think that FK stands for Fricke and Kopff, rather than for Fundamental Katalog. As a result of this fundamental work, he began to concentrate on the proper motion systems and the constant of precession. He discussed the southern hemisphere problems where the knowledge of stellar motions was considerably less well developed than in the northern hemisphere and urged southern hemisphere observatories to do fundamental astrometry. Alas, his urgings were not successful. However, around that time, many countries joined to establish the southern reference star system, sent expeditions to the southern hemisphere, and enlisted telescopes at existing observatories.

In an article in Vistas in 1965, Fricke pointed out that people often think the derivation of statistical parallaxes can be accomplished without great effort. "That is a misconception," he said. I would think that this was a typical understatement. He proceeded to show in this article, through comparison of the GC, N30 and FK 4 systems, that for stars at distances less than 100 parsec, effects of differences in the systems are negligible; for stars at a few 100 parsec they are appreciable; and for distances greater than 1000 parsec the effects are disastrous. For example, he found that where N30 gave 2000 parsec for a group of stars, FK 4 gave 1000 parsec for the same group.

Fricke's influence on the field of fundamental astronomy now became intense. He bombarded the world with invited talks, with review papers, and with a very considerable series of fundamental discussions on the precession constant, its influence and reasons for replacing Newcomb's constant (which was not done in the 1964 IAU round of changes). He got right in the thick of the arguments by radio astronomers who, after having invented VLBI were saying that they would measure the precession constant more accurately in just a few years. Fricke argued that we needed it then, not a few years later (and in fact doubted that it could be obtained in a few years - and he was right!) because of the need to transfer the fundamental reference system into a dynamical inertial system. In order to do that, we need both the FK 5 and a new constant of precession at the same time. In 1976 he made the categorical statement that if we wait with precession then we must wait with the FK 5 and stay with the FK 4 until the radio astronomers have produced results.

We are now getting closer to the introduction of the FK 5 system and in particular, the question of the motion of the equinox, which had to be resolved. I very vividly remember, with glee, the Montreal IAU meeting in 1979. At a joint meeting of several Commissions where the equinox motion and other matters were discussed, Fricke wrote his formula for E on the board. After some discussion and some eloquent statement by Fricke about the quality of the data used for this, it was adopted as an IAU value. More discussion ensued and during this discussion it became clear that this was probably not the final value! Fricke sounded somewhat amazed that people did not realize that he required another year to come up with a final number. However,the confidence in Fricke was now so great that, although very reluctantly, he was left to determine the final value in the next year or so.

I cannot resist the temptation of joking a little bit on the preparation of the FK 5. In 1960, in a paper in the "Astronomical

Journal" Fricke stated that "it can be assured that a future revision of FK 4 using new data should no longer be a major undertaking because of the use of modern computer techniques." The FK 3 was completed in 1935. The FK 4 was started in 1952 and completed in 1963; thus it took 11 years. Work on the FK 5 was started in about 1965 and presumably the FK 5 will be ready in 1986; thus it will take 21 years to complete the FK 5 with modern computers - "no longer a major undertaking!?" In the same paper in 1960, Fricke remarked "no final results can be presented at this moment for the systematic correction from FK 3 to FK 4." In the last few years, we have heard several times exactly the same statement about the conversion from FK 4 to FK 5.

Fricke has high hopes for the interconnection of the radio and optical reference frames through the work with the Hipparcos satellite and the Space Telescope. He does not have much faith in tying the reference frame to galaxies because accuracy can be obtained with other means. In this respect, in 1977 Fricke told us that we should not worry about the accuracy of the precession constant, because: "when single observations of position can be made with an accuracy of better than 0".03 (systematic and individual), a whole new field will be opened for the discovery of a variety of small effects of great physical importance (examples are the nonrigidity of the Earth and relativistic effects in dynamics)." It seems to me that we are close to reaching that exciting era, and we are getting there in a systematic and well thought-out way with a sense of historical perspective, rather than by helter-skelter means. Our thanks are due to Walker Fricke whose tremendous expertise and drive have brought the field of fundamental astronomy to its current state.

ANNOUNCEMENT

Fundamental Astronomy and Solar System Dynamics

Invited Papers Honoring Prof. Walter Fricke
on the Occasion of His 70th Birthday

Edited by R. L. DUNCOMBE, J. H. LIESKE, and P. K. SEIDELMANN

Please note that a hardbound edition of this special issue of *Celestial Mechanics*, Vol. 37, No. 3 (November 1985) is available from the publishers.

ISBN 90–277–2268–4 Prices: Dfl. 100,– / US$ 47.– / £ 31.95

GPSR Compliance

The European Union's (EU) General Product Safety Regulation (GPSR) is a set of rules that requires consumer products to be safe and our obligations to ensure this.

If you have any concerns about our products, you can contact us on ProductSafety@springernature.com

In case Publisher is established outside the EU, the EU authorized representative is:

Springer Nature Customer Service Center GmbH
Europaplatz 3
69115 Heidelberg, Germany

Batch number: 10584568

Printed by Printforce, the Netherlands